Friendship 7

The First Flight Of John Glenn

The NASA Mission Reports

Compiled from the NASA archives & Edited
by Robert Godwin

All rights reserved under article two of the Berne Copyright Convention (1971).
We acknowledge the financial support of the Government of Canada through the
Book Publishing Industry Development Program for our publishing activities.
Published by Apogee Books an imprint of Collector's Guide Publishing Inc., Box 62034, Burlington, Ontario, Canada, L7R 4K2
Printed and bound in Canada by Webcom Ltd of Toronto
Friendship 7 — The First Flight Of John Glenn - The NASA Mission Reports
Edited by Robert Godwin
ISBN 1-896522-60-2
All photos courtesy of NASA

Introduction

In October 1998, the space shuttle lifted off from Cape Canaveral carrying John Glenn on his second trip into Earth orbit. After months of media hype the veteran astronaut was finally getting his hard-earned wish.

If you were to scrutinize the press over the weeks and months leading up to the flight you would be hard pressed to find an article which didn't at some point raise doubts as to whether Glenn's return was good science or just some sort of payback for an influential Senator.

I myself found that some arguments raised by the pundits were often convincing, and at times I wondered how much the flight was actually going to achieve in the way of hard results for the life-sciences. However, I have now firmly changed my mind.

While editing this book I learned a lot about the kind of person John Glenn must be. Tom Wolfe encapsulated this special character in his now legendary book "The Right Stuff" — but just what does it mean to have the Right Stuff? In this book I hope that you the reader will find out — as I did.

In February 1962 John Glenn would be the third man to be pushed into space in the Mercury spacecraft, however he would be the first man to go into orbit under the scrutiny of the world media, and he would be the first to do so riding atop the Atlas missile.

Although Shepard and Grissom had already ventured into the unknown riding up on the top of the much smaller Redstone rocket, their flights only lasted 15 minutes and they rode a trajectory much like a cannon shell firing out of a gun. The total time that either would be subjected to weightlessness was something under four minutes. Meanwhile the Russians had orbited Gagarin but were not noted for sharing data. Gagarin's flight was still almost a total mystery in the West.

When John Glenn strapped into his capsule, dubbed Friendship 7, the experts at NASA were sending him into the unknown. The most fundamental questions were unanswered. Would he be able to see, or would the lenses of his eyes not function in zero-g? Would he be able to swallow or would he choke on his food? Even the most elementary and embarrassing questions were still unanswered. Would his bladder function normally?

John Glenn had to climb into a vehicle the size of a telephone booth, probes and cuffs attached to every conceivable part of his body, and then allow himself to be shut in with a little more than 24 hours worth of air and wait for someone to ignite the engines on a booster that had until very recently been more noteworthy for it's cataclysmic failures than for it's successes. Once the vehicle left the launch pad Glenn would be subjected to no less than 7.7 times gravity, a force which would make moving his arms all but impossible. Once in orbit, Glenn then was expected to spend the next 4¾ hours conducting every conceivable test that could be fitted into the schedule. Thankfully he was spared such charming procedures as puncturing his own space-suit with a hypodermic and injecting himself with anti-nausea drugs.

Little time was left for sight-seeing, and the capsule was so confined that the more pleasant aspects of zero-g were mostly left to the imagination. After only an hour or so, the spacecraft's automatic guidance system failed and Glenn was then obliged to fly the vehicle manually, which required the cancellation of some tests as he wrestled with the problems of making sure he could fly his ship home. Then, as the final orbit began, a light illuminated on someone's console in Houston and told the flight controllers that the inflatable landing bag on Friendship 7 had deployed. If this was true then the heat-shield, the only protection against the scorching heat of reentry, was probably loose.

When Glenn was notified of this, only minutes before he was due to return to Earth, his reaction was almost as though this was just a minor problem. If you read the transcript of the conversation between mission control and Glenn, he calmly asks why he is ordered to leave his Retro-pack in place during re-entry. When the explanation comes he replies with nothing more than," Roger, understand. I will have to make a manual 0.05g entry when it occurs, and bring the scope in manually. Is that affirm?" The jargon continues unabated as though he is running through his grocery list.

Once his re-entry cycle begins, Glenn is surrounded by a fireball of debris and ionized gas as the retro-pack splinters away, every second must have seemed an eternity as he waited to see if he would survive the inferno. As the parachutes and snorkels deployed the intense heat of re-entry finally began to seep into the tiny capsule and the cabin temperature peaked at 103°F while he calmly sat on the ocean waiting to be picked up by the recovery helicopter. With only a couple of minor comments about the painful heat and humidity inside his rubberized space suit, Glenn finally walked onto the deck of the USS Noa and into retirement from space flight.

Before the flight of Friendship 7 Glenn and his six companion astronauts were subjected to criticism for being over-prepared. After the flight a new template had been molded. A template from which the "Right Stuff" could sometimes be fashioned. Before the Mercury Seven and before John Glenn, there were no astronauts, there was no yardstick against which to gauge the expected behavior of an astronaut. John Glenn gave all that followed him something to aim for. His discipline and dedication were unparalleled, his demeanor as a spokesperson after his flight was inspirational. His physical endurance and patience were not much short of astonishing. Glenn allowed himself to be put into the line of fire for everyone who has followed him. He and his fellow Mercury astronauts *explored,* and they showed us all that perhaps our path into the heavens is dangerous but also attainable.

John Glenn exhibited to the world that he was the perfect choice to fly the first orbital flight, he was healthy, an expert pilot, cool under duress, a compliant Guinea Pig, intelligent and cooperative. These are qualities not often found — much less in one individual.

This brings me back to 1998. If NASA is going to conduct gerontology experiments in orbit, who is more eminently qualified to be the Guinea Pig than John Glenn. When he was in his forties his medical status was put under a microscope (in fact sizable chunks of it are in this book). He was in better shape in 1998 than most people twenty years his junior, and he was experienced in the process of going into space.

In the final analysis the hacks never questioned the validity of conducting experiments on aging, they only debated whether Glenn should be the test subject and whether any useful science would be done. Those decisions were made back in 1961 when he was picked for flight MA-6. The professional manner with which John Glenn has conducted his entire career from the military, to NASA and the US Senate, is clear evidence that he wouldn't waste a single precious minute of his time in orbit and if anyone could get meaningful data it was him, and besides — better the daredevil you know.

Robert Godwin
(Editor)

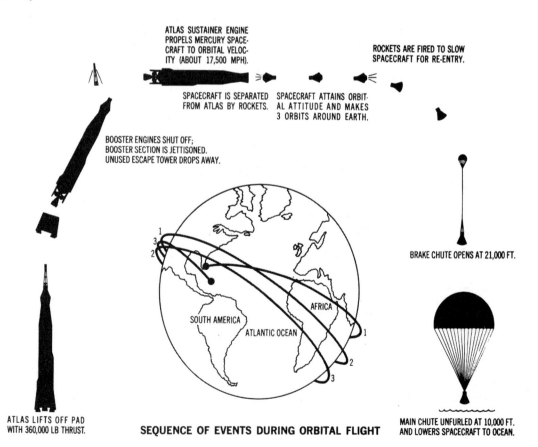

SEQUENCE OF EVENTS DURING ORBITAL FLIGHT

Friendship 7
The First Flight Of John Glenn
The NASA Mission Reports

(from the archives of the National Aeronautics and Space Administration)

Contents

Astronaut John H. Glenn Jr. walking out of hangar with Dr William Douglas and suit technician Joe Schmitt

Post flight parade in Cocoa Beach with President John Kennedy and General Leighton Davis

NEWS RELEASE

FOR RELEASE: Sunday a.m.'s
January 21, 1962

Release No. 62-8

MERCURY - ATLAS 6 AT A GLANCE

MISSION — Manned orbital flight to (1) evaluate the performance of a man-spacecraft system; (2) investigate man's capabilities in the space environment; (3) obtain the pilot's opinions on the operational suitability of the spacecraft and supporting systems for manned space flight.

LAUNCH DATE — The flight currently is scheduled no earlier than January 24, 1962. On whatever day, the launch will be attempted between 7:30 a.m. and 12:30 p.m. and may "slip" on a day-to-day basis as required. Launch timing will be planned to provide at least three hours of daylight search time in the probable recovery areas.

FLIGHT DURATION — Depending on literally thousands of variables, the Mercury Operations Director (Manned Spacecraft Center Associate Director Walter C. Williams) may elect a one, two or three-orbit mission. That decision will be made only minutes before launch and may be changed at any time during the mission. Recovery after one full orbit is planned for about 500 miles east of Bermuda; after two orbits, some 500 miles south of Bermuda; three orbits, about 800 miles southeast of Cape Canaveral, Fla. Each orbit takes about 90 minutes, carrying the craft between 100 and 150 miles altitude, 32 degrees north and south of the equator.

If the mission ends after orbit one or two, the astronaut will be moved to the Kindley Air Force Base Hospital in Bermuda for a 48- hour rest and debriefing. If the mission goes a full three orbits, he will be flown to Grand Turk Island (Bahamas) for a similar operation before being returned to the mainland.

PILOT — Astronaut John H. Glenn, Jr., 40. A lieutenant colonel In the United States Marine Corps, Glenn has been with NASA for three years on a detached duty basis. Backup pilot for this flight is Astronaut M. Scott Carpenter, 36. (See biographies).

SPACECRAFT — Bell-shaped, the MA-6 craft — listed as No. 13 in engineering documents and named "Friendship 7" by Astronaut Glenn — stands 9 feet high and measures 6 feet across the base. Spacecraft weight at launch will be about 4,200 pounds; spacecraft weight in orbit (after jettisoning of escape tower) — 3,000 pounds; on-the-water recovery weight — 2,400 pounds. Prime contractor for the spacecraft is McDonnell Aircraft Corp. of St. Louis, Mo.

LAUNCH VEHICLE — A modified Atlas D is used to launch orbital Mercury missions , reaching a speed of 17,500 miles per hour. At launch, booster and spacecraft stand 93 feet tall, including a 16-foot tower above the spacecraft. The tower contains a solid propellant rocket hooked to an abort sensing system. Should trouble develop on the launch pad or in the early boost phase of the mission, the escape system will be triggered automatically or by the pilot or from the ground to pull the spacecraft away from the booster. The booster is manufactured by the Astronautics Division of General Dynamics Corp.

NETWORK — The Mercury Tracking Network consists of 18 stations around the world, including two ships, one on the equator in the Atlantic off the coast of Africa and the other in the Indian Ocean. Some 500 technicians man these stations, all of which are in radio or cable communication with the Mercury Control Center at the Cape via the NASA Goddard Spaceflight Center at Greenbelt, Md.

RECOVERY — More than 20 ships will be deployed in the Atlantic alone to take care of prime and contingency recovery areas. Recovery forces are under the command of Rear Admiral John L. Chew, Commander of Destroyer Flotilla Four. In addition, ships and rescue planes around the world will go into action in the event of an emergency landing. More than 15,000 men will have a hand in the recovery, search and rescue effort.

RESPONSIBILITIES — Project Mercury, the nation's first manned space flight research project, was conceived and is directed by the National Aeronautics and Space Administration, a civilian agency of the government charged with the exploration of space for peaceful and scientific purposes. Technical project direction for Mercury is supplied by NASA's Manned Spacecraft Center, directed by Robert R. Gilruth at Langley Field, Va., and soon to move to Houston, Texas. The Department of Defense, largely through the Air Force and Navy, provides vital support for Mercury. DOD support is directed by Major General Leighton I. Davis, USAF, Commander of the Atlantic Missile Range. In all, more than 30,000 persons will have a part in this mission, including government and industry.

PROJECT COST — Total Project Mercury cost through orbital flight is estimated at $400 million. About $160 million will have gone to the prime spacecraft contractor, McDonnell and its subcontractors and suppliers; $95 million for the network operations; $85 million for boosters, including Atlases, Redstones and Little Joes; $25 million for recovery operations and roughly $35 million for supporting research in diverse areas.

MISSION PILOT TASKS — The MA-6 pilot will perform many control tasks during flight to obtain maximum data on spacecraft performance, his own reactions to weightlessness and stress, and to study the characteristics of the earth and stars from his vantage point over 100 miles above the earth's surface.

The Astronaut will participate actively during the flight. This will include the following tasks:

(1) Manage the operation of all spacecraft systems, particularly the attitude control system, electrical system, environmental control system, and communications systems.

(2) Observe and correct any discrepancies in system operation. Discrepancies will be correlated with telemetered observations received at ground stations.

(3) Monitor critical events during launch, and terminate the mission if necessary.

(4) Maintain a complete navigation log during flight which will enable him to compute his retro-fire time if ground communications should fail. This onboard navigation will include periscope ground sightings which indicate position over ground and altitude.

(5) Ground communications to receive updated retro-fire information, and receive detailed behavior of spacecraft systems as determined from ground telemetry.

(6) Evaluate his physical condition to augment the biomedical data which are telemetered to the ground.

About every 30 minutes, the pilot will make detailed voice reports on spacecraft systems and operations conditions. His own transmissions will include critical information as mode of control, precise attitude, planned retro-fire time, control system fuel, oxygen, and coolant.

MISSION PROFILE

POWERED FLIGHT — The manned Mercury spacecraft will be launched atop an Atlas from Cape Canaveral following a two-day split countdown. Technical conditions or weather could, of course, delay the launch from minutes to days.

According to the flight plan, the spacecraft will be launched on a path along the Project Mercury World-wide Tracking Range on a launch heading of about 72 degrees just north of east from Cape Canaveral.

An internal programmer in the Atlas will guide the vehicle from liftoff until staging occurs. All of the Atlas liquid propellant engines are ignited before lift-off.

At staging, about two minutes after lift-off, the two booster engines will drop off and the sustainer and vernier engines will continue to accelerate the vehicle. Staging occurs at an altitude of about 40 miles and a range of about 45 miles from the launch pad.

During the first 2½ minutes of flight, an electronic brain, called the Abort Sensing and Implementation System (ASIS) is capable of sensing impending trouble in the rocket and triggering the escape rocket. The astronaut can also trigger the Mercury escape rocket to pull the spacecraft away from the Atlas.

About 20 seconds after staging, and assuming the flight is proceeding as planned, the 16-foot escape tower and rocket will be jettisoned. Landing systems will be armed. The Mercury-Atlas vehicle will continue to accelerate toward the orbit point guided by ground command guidance.

Until orbital insertion, the abort sensing system will continue to watch for trouble. If significant deviation should occur, the system will actuate circuits to release the spacecraft-to-Atlas clamp ring and fire the posigrade rockets on the base of the spacecraft.

About five minutes after lift-off, guidance ground command will shut down the sustainer and vernier engines. As the engines shut down, the spacecraft-to-booster clamp ring is released automatically and posigrade rockets are fired to separate the craft from the Atlas.

ORBITAL INSERTION — After a few seconds of automatic damping — getting rid of any unusual motions — the spacecraft will swing 180 degrees so that the blunt face of the craft is turned forward and upward 34 degrees above the horizontal. From that point on during orbital flight, the spacecraft can be controlled in proper attitude automatically or manually by the pilot.

If all goes well, the Mercury spacecraft will be inserted into orbit in the vicinity of Bermuda. By that time the vehicle will be at an altitude of approximately 100 miles and traveling at a speed of about 17,500 miles per hour. At engine cut-off the craft will have been subjected to more than 7½ "G". Reentry "G" will also reach 7½.

A three-orbit flight will last approximately 4¾ hours; a two-orbit flight, 3¼ hours; one orbit, 1 ¾ hours. The Mercury craft will reach a peak altitude (apogee) of about 150 statute miles off the West Coast of Australia and a low point (perigee) of 100 miles, at the insertion point near Bermuda.

REENTRY — After the desired number of orbits, as the spacecraft approaches the West Coast of North America, retro or braking rockets will be fired to initiate reentry.

Shortly after the retrorockets are fired, the exhausted rocket package will be jettisoned and the spacecraft automatically will assume reentry attitude. "Friendship 7" will begin to encounter more dense atmosphere of the Earth approximately over the east coast at an altitude of about 55 miles. At this point, temperatures will start mounting on the spacecraft's ablation heat shield.

On a nominal mission, peak reentry temperature of about 3,000 degrees F. will occur at 25 miles altitude while the spacecraft is moving at nearly 15,000 miles per hour. All told, the craft will sustain temperatures in this neighborhood for about two minutes.

Almost coincident with the heat pulse is a dramatic reduction in capsule speed. Between 55 miles and 12 miles altitude covering a distance of 760 miles - spacecraft velocity should go from 17,500 miles per hour down to 270 miles per hour in a little over five minutes.

At about 21,000 feet, a six-foot diameter drogue parachute will be opened to stabilize the craft. At about 10,000 feet, a 63-foot main landing parachute will unfurl from the neck of the craft.

On touchdown, the main chute will be jettisoned. On-board electrical equipment will be shut down, and location aides will be activated.

RECOVERY — Several new recovery techniques will be tried operationally for the first time in MA-6. If all's well, the astronaut, on leaving the craft via the neck or the side hatch, should be greeted by two frogmen who by that time will have cinched a new flotation belt around the base of the craft. This is to add to the craft's seaworthiness.

Plans call for the frogmen to leap into the water with the quick-inflating belt from one of several recovery helicopters staged off aircraft carriers in the three prime recovery zones. As soon as they have secured the three foot high belt, the astronaut will emerge, grab a "horse -collar" lift from a hovering helicopter and be whisked up into the copter and to a nearby carrier.

Meanwhile, a smaller ship is to go along side the spacecraft and hoist it onto its deck for transfer to the carrier or direct delivery to the Cape.

THE MA-6 MERCURY SPACECRAFT

The MA-6 spacecraft is similar to those used in previous Mercury flights. About as big as a phone booth, the interior looks much like the cockpit of a high-performance airplane only smaller.

Here are some of the major items confronting the pilot:

Instrument Panel — Instruments are located on a main instrument panel, a left console, and a right console. The main panel is directly in front of the pilot. Navigational and control instruments are located in the left and center sections of the panel and the periscope is located in the center. The right section of the main panel is composed of environmental system gauges and controls, electrical switches, indicators and communication system controls.

Rate Stabilization and Control — Attitude of the Mercury spacecraft is changed by the release of short bursts of superheated steam (hydrogen peroxide) from 18 thrust nozzles located on the conical and cylindrical portions of the craft's surface. Timing and force of these bursts are controlled by one of the following: (1) Automatic Stabilization and Control System (ASCS) or "Auto-Pilot"; (2) Rate Stabilization and Control System (RSCS), or "Rate Command" System (3) the Manual Proportional Control System, a manual-mechanical system, and (4) the Fly-By-Wire (FBW), a manual-electrical system.

The left console includes sequencing telelights and a warning panel, indicators and controls for the spacecraft's automatic pilot (ASCS), environmental control and landing systems. Altogether, there are over 100 lights, fuses, switches and miscellaneous controls and displays.

Cameras — A 16mm camera is installed to the left of the astronaut's head to photograph the instrument panel display from launch through recovery. A pilot observer camera is mounted in the main instrument panel and will also be operated from launch through recovery.

Periscope —An earth periscope is located approximately two feet in front of the pilot and will provide a 360-degree view of the horizon. The pilot may manually adjust for "low" or "high" magnification. On "low" he will have a view of the Earth of about 1,900 miles in diameter — in "high", the field of view will be reduced to about 80 miles. Altitude can be measured within plus or minus 10 nautical miles. The Mercury Earth periscope will, in addition, serve as a navigational aid.

Pilot Support Couch — The astronaut's couch is constructed of a crushable honeycomb material bonded to a fiberglass shell and lined with rubber padding. Each astronaut has a flight couch contoured to his specific shape. The couch is designed to support the pilot's body loads during all phases of the flight and to protect him from acceleration forces of launch and reentry.

Restraint System — The restraint system, which consists of shoulder and a chest strap, leg straps, crotch strap, lap belt and toe guards, is designed to keep the astronaut in the couch during maximum deceleration.

Environmental Control System — The environmental control system provides the MA-6 spacecraft cabin and the astronaut with a 100-percent oxygen environment to furnish breathing, ventilation, and pressurization gas required during flight. The system is completely automatic, but in the event the automatic control fails, emergency controls can be used. The system consists of two individual control circuits (the cabin circuit and the suit circuit), which will normally operate for about 28 hours. Both systems are operated simultaneously. The suit circuit is isolated from the cabin circuit by the astronaut closing the face-plate on his helmet. Unless there is a failure in the cabin circuit causing loss of pressure, the pilot's pressure suit will not be inflated.

Aeromedical Information — Throughout the flight, the physical well-being of the pilot will be monitored. The pilot's respiration rate and depth, electrocardiogram and body temperature will be telemetered to flight surgeons on the ground.

Pilot Communications — The MA-6 astronaut may remain in touch with the ground through the use of high-frequency and ultra-high-frequency radios, radar recovery beacons, and if the situation dictates, a command receiver or a telegraph code key.

Main battery system — Three 3,000 watt-hour batteries and one 1,500 watt-hour battery are connected in parallel to provide power for the complete mission and about a 12 hour post-landing period. A standby backup power system of 1,500 watt-hour capacity is also provided. To further insure reliable operation of the pyrotechnic system, each device has a completely isolated power feed system.

Retrofire Timer — There will be a timer in the MA-6 spacecraft with three major separate operational components, (1) a standard aircraft elapsed time clock, (2) a "seconds from launch" digital indicator with a manual reset, and (3) a resettable timer and time-delay relay which will initiate the retrograde fire sequence. When the preset time has passed, the relay closes and actuates the retrograde fire signal, at the same time sending a telemetered signal to the ground.

Altimeter — The Mercury barometer altimeter is a single revolution indicator with a range from sea level to 100,000 feet. The dial face has reference marks at the drogue and main parachute deployment altitude. At the top right corner of the main panel are located environmental displays, providing the pilot with indications of cabin pressure, temperature, humidity, and oxygen quantity remaining.

Food, Water, and Waste Storage — "Friendship 7" will carry about 3,000 calories of food — beef and mixed vegetables — and about six pounds of water. The water will be in two flat bottles, each fitted with a tube. The food is in two tubes, about the size of tooth paste tubes. In addition, there will be some quick-energy sugar tablets.

Survival -Equipment — The survival package will consist of a one man life raft, desalting kit, shark repellant, dye markers, first aid kit, distress signals, a signal mirror, portable radio, survival rations, matches, a whistle, and ten feet of nylon cord. A new lightweight, radar-reflective life raft is fabricated of Mylar (for air retention) and nylon (for strength). The three-pound, four-ounce raft features three water ballast buckets for flotation stability and a deflatable boarding end which may be reinflated by an oral inflation tube following boarding. The raft, made of the same material used in the Echo satellite balloon, is international orange.

Pilot's Map — A small cardboard diagram of the MA-6 flight path with recovery forces indicated is contained within a bag suspended beneath the periscope. On the reverse side, the pilot's view through the periscope from maximum altitude is shown. Last minute information on cloud formations and weather phenomena will be marked by Mercury weather experts.

Hatch — The MA-6 spacecraft is equipped with a hatch secured by explosive bolts, just as the pilot's canopy is secured in a high performance aircraft. The astronaut can jettison the hatch by pushing a plunger button inside the spacecraft or by pulling a cable. The hatch may also be removed by recovery teams.

Cylindrical Neck Contents — Above the astronaut's cabin, the cylindrical neck section contains the main and reserve parachute chute system.

Three parachutes are installed in the spacecraft. The drogue chute has a six-foot-diameter, conical, ribbon-type canopy with approximately six foot long ribbon suspension lines, and a 30-foot-long riser made of dacron to minimize elasticity effects during deployment of the drogue at an altitude of 21,000 feet. The drogue riser is permanently attached to the capsule antenna by a three point suspension system terminating at the antenna in three steel cables, which are insulated in areas exposed to heat.

The drogue parachute is packed in a protective bag and stowed in the drogue mortar tube on top of a light-weight sabot. The sabot functions as a piston to eject the parachute pack, when pressured from below by gasses generated by a pyrotechnic charge.

The function of the drogue chute is to provide a backup stabilization device for the spacecraft in the event of failure of the Reaction Control and Stabilization System. Additionally, the drogue chute will serve to slow the capsule to approximately 250 feet per second at the 10,000 foot altitude of main parachute deployment.

The reserve chute is identical to the main chute. It is deployed by a flat circular-type pilot chute.

Other components of the landing system include drogue motor and cartridge, radar chaff, barostats, antenna fairing ejector, sea marker packet, and two underwater charges — one attached to and deployed by the antenna lanyard and the other fixed to the spacecraft.

Following escape tower separation in flight, the 21,000 and 10,000 foot barostats are armed. No further action occurs until the capsule descent causes the 21,000 foot barostat to close and activate power to the drogue mortar which ejects the drogue as well as the radar chaff — loosely packed under the drogue chute — to provide a target for radar location.

Two seconds after the 10,000 foot barostat closes, power is supplied to the antenna fairing ejector — located above the cylindrical neck section — to deploy the main landing parachute and an underwater charge, which is dropped to provide an audible sound landing point indication. The ultra-high frequency SARAH radio then begins transmitting. A can of seamarker dye is deployed with the reserve chute and remains attached to the spacecraft by a lanyard.

On landing an impact switch jettisons the landing parachute and initiates the remaining location and recovery aids. This includes release of sea-marker dye with the reserve parachute if it has not previously been deployed triggering a high-intensity flashing light, extension of a 16-foot whip antenna and the initiation of the operation of a high-frequency radio beacon.

If after landing, the spacecraft should spring a leak or if the life support system should become fouled after landing, the astronaut can escape through this upper neck section or through the side hatch.

Impact Skirt — Following deployment of the main landing parachute, the heat shield is released, extending the landing impact bag to form a pneumatic cushion primarily for impact on land, it is also required for capsule stability after water landings.

The air cushion is formed by a four-foot skirt made of rubberized fiberglass that connects the heat shield and the rest of the capsule. After the main parachute is deployed, the heat-shield is released from the capsule and the bag fills with air. Upon impact, air trapped between the heat shield and the capsule is vented through holes in the skirt as well as portions of the capsule which are not completely air tight, thereby providing the desired cushioning effect.

THE ATLAS LAUNCH VEHICLE

The launch vehicle to be used for the Mercury-Atlas 6 test is an Atlas D model, one of several Atlases especially modified for use in the Mercury flight test program. This vehicle develops 360,000 pounds of thrust and burns Rp-1, a kerosene-like fuel, and liquid oxygen.

Principle differences in the Mercury-Atlas and the military version of the vehicle include: (1) modification of the payload adapter section to accommodate the Project Mercury spacecraft, (2) structural strengthening of the upper neck of the Atlas to provide for the increase in aerodynamic stress imposed on the Atlas when used for Mercury missions, and (3) inclusion of an automatic Abort Sensing and Implementation System (ASIS) designed to sense deviations in the performance of the Atlas and trigger the Mercury Escape System before an impending catastrophic failure.

The Atlas measures 65 feet from its base to the Mercury adapter section and is 10 feet in diameter at the tank section. With adapter section, spacecraft and escape tower, the Mercury-Atlas stands 93 feet tall.

The Atlas is constructed of thin-gage metal and maintains structural rigidity through pressurization of its fuel tanks. For this flight, the Atlas will have a heavier gage skin at the forward end of the liquid oxygen tank, the same as that used in other launches of Atlas space systems.

All five engines are ignited at the time of launch - the sustainer (60,000 pounds thrust), the two booster engines (150,000 pounds thrust each) which are outboard of the sustainer at the base of the vehicle and two small vernier engines which are used for minor course corrections during powered flight. During the first minute of flight, it consumes more fuel than a commercial jet airliner during a transcontinental run.

ASTRONAUT PARTICIPATION

All seven of Project Mercury's team of astronauts will participate in the MA-6 orbital mission, some as flight controllers from far flung vantage points around the globe.

Astronauts John H. Glenn, Jr., prime pilot, M. Scott Carpenter, backup pilot, and Alan B. Shepard, Jr., technical advisor, will be at Cape Canaveral. Astronaut Walter M. Schirra, Jr. will be stationed at the Mercury site at Pt. Arguello, California, and Astronaut L. Gordon Cooper, Jr. will participate from the Mercury tracking station in Muchea, Australia. Astronaut Virgil I. "Gus" Grissom will monitor launch, insertion, landing, and recovery from Mercury's Bermuda station, while Donald K. Slayton will perform spacecraft checkout prior to insertion of the mission pilot.

THE NETWORK

World-wide Mercury tracking stations, including ships in the Indian and Atlantic Oceans, will monitor the MA-6 flight. The Space Computing Center of the NASA Goddard Space Flight Center in Greenbelt, Maryland, will make trajectory computations.

During the flight, information will pour into the Space Computing Center from tracking and ground instrumentation points around the globe at the rate, in some cases, of more than 1,000 bits per second. Upon almost instantaneous analysis, the information will be relayed to the Cape for action.

In addition to proving man's capability for surviving in and performing efficiently in space — and since only two other flights have been made along the world-wide tracking network — the test will further evaluate the capability of the network to perform tracking, data-gathering and flight control functions.

The Mercury network, because of the man factor, demands more than any other tracking system. Mercury missions require instantaneous communication.

Tracking and telemetered data must be collected, processed, and acted upon in as near "real" time as possible. The position of the vehicle must be known continuously from the moment of lift-off.

After injection of the Mercury spacecraft into orbit, orbital elements must be computed and prediction of "look" information passed to the next tracking site so the station can acquire the capsule.

Data on the numerous capsule systems must be sent back to Earth and presented in near actual time to observers at various stations. And during the recovery phase, spacecraft impact location predictions will have to be continuously revised and relayed to recovery forces,

An industrial team headed by Western Electric Company recently turned over this $60 million global network to the National Aeronautics and Space Administration.

Other team members are Bell Telephone Laboratories, Inc.; the Bendix Corporation; Burns and Roe, Inc and International Business Machines Corporation. At the same time, the Lincoln Laboratory of Massachusetts Institute of Technology also has advised and assisted NASA on special technical problems relating to the network.

Concluding the contract involved extensive negotiations with Federal agencies, private industry, and representatives of several foreign countries in the establishment of tracking and ground instrumentation.

The network consists of 18 stations. The system spans three continents and three oceans, interconnected by a global communications network. It utilizes land lines, undersea cables and radio circuits, and special communications equipment installed at commercial switching stations in both the Eastern and Western hemispheres.

The project includes buildings, computer programming communications and electronic equipment, and related support facilities required to direct, monitor, and provide contact with the nation's orbiting Project Mercury astronaut.

Altogether, the Mercury system involves approximately 60,000 route miles of communications facilities to assure an integrated network with world-wide capability for handling satellite data. It comprises 140,000 actual circuit miles 100,000 miles of teletype, 35,000 miles of telephones, and over 5,000 miles of high-speed data circuits.

Sites linked across the Atlantic are: Cape Canaveral, Grand Bahama Island, Grand Turk Island, Bermuda, Grand Canary Island, and a specially fitted ship in Mid-Atlantic.

Other stations in the continental United States are at Ft. Arguello in Southern California; White Sands, New Mexico; Corpus Christi, Texas; and Eglin, Florida. One station is located on Kauai Island in Hawaii.

There are two radar picket ships. The Atlantic Ship Rose Knot will be stationed on the equator near the West African coast The Indian Ocean Ship Coastal Sentry will be located midway between Zanzibar and Muchea, Australia.

Stations at overseas sites include one on the south side of Grand Canary Island, 120 miles west of the African Coast; Kano Nigeria, in a farming area about 700 rail miles inland; Zanzibar an island 12 miles off the African coast in the Indian Ocean; two in Australia - one about 40 miles north in Perth, near Muchea, and the other near Woomera; Canton Island, a small coral atoll about halfway between Hawaii and Australia; one In Mexico, near Guaymas on the shore of the Gulf of Mexico; and one in Bermuda, an independent, secondary control center.

Some 20 private and public communications agencies throughout the world provided leased land lines and overseas radio and cable facilities.

Site facilities include equipment for acquiring the spacecraft; long range radars for automatic tracking; telemetry equipment for receiving data on the spacecraft and the astronaut; command control equipment for controlling the manned vehicle from the ground, if necessary, and voice channels for ground-to-air communications. The extensive ground communications systeminterconnects all stations through Goddard and the Mercury Control at Cape Canaveral.

Sites equipped with tracking radars have digital data conversion and processing equipment for preparing and transmitting information to the computing system without manual processing, marking a significant achievement — global handling of data on a real-time basis.

One function of the computer is to transmit information regarding the spacecraft's position to Mercury Control Center at the Cape, where it is displayed on the world map in the Operations Room. The computer also originates acquisition information which is automatically sent to the range stations.

During every major Mercury launch, the attention of some 15 NASA flight controllers is focused on dozens of consoles and wall displays in the Mercury Control Center Operations Room. This room is the control point for all information that will flow through the world-wide tracking and communications system. In this room NASA Flight Controllers make all vital decisions required, and issue or delegate all commands.

In the fifty-foot square room, about 100 types of information register at various times on the indicators of the consoles and the high range-status map of these 100 quantities, 10 show biomedical condition, approximately 30 relate to life support facilities and about 60 give readings on spacecraft equipment. This information flows in on high-speed data circuits from computers at the Goddard Center, on direct teletype circuits from remote sites, and by booster and spacecraft telemetry relayed over radio and wire circuits.

Three kinds of data start pouring into the computing system as soon as the booster lifts half an inch off the launch pad;

1. Radar data: Triggers the Cape Canaveral IBM 7090 which monitors the spacecraft's flight path and predicts its impact point if the mission must be aborted.

2. Guidance data: Radioed from the spacecraft to a special purpose computer at the Cape.

3. Telemetry data: Reports check points, e.g., liftoff, booster separation.

These data are transmitted from Cape Canaveral to Goddard where IBM 7090's compare the spacecraft trajectory to a predetermined flight path — and flash the results back to Canaveral. This is a "real time" operation — that is, the system receives, moves it over 2,000 miles, analyzes, predicts and displays data so fast that observers and controllers follow events as they happen.

COMMUNICATIONS INFORMATION

The system carries telephone, teletype and high-speed data (1,000 bits per second) information. It can accept a message from a distant site and delivers it to the final destination regardless of location - in a little over one second.

Radio teletype facilities use single sideband transmitters, which are less susceptible to atmospheric interference. All circuits, frequencies and paths were selected only after a careful study of data accumulated over 25 years by the National Bureau of Standards on the various propagation qualities of many radio paths.

Submarine cables to London (via New York), to Hawaii (via San Francisco), and to Australia- (via Vancouver, B.C.) are included in the Mercury communications network.

The Mercury Voice Network has a twofold mission:

1. Provide Mercury Control Center (MCC) with "real time" information from world-wide tracking stations having the orbiting Mercury spacecraft in view.

2. Provide a rapid means for dealing with emergency situations between MCC and range stations during a mission.

The network is essentially a private line telephone system radiating from GSFC to MCC and the project's world sites.

These lines are used during an orbit mission to exchange verbal information more rapidly than can be done by teletype. Conversations are recorded both at Goddard and Mercury Control Center for subsequent playback. When not used for orbit exercise the circuits are utilized for normal communications operations.

ASTRONAUT TRAINING PROGRAM SUMMARY

Here are some of the general training activities that the Project Mercury astronauts have undergone since May, 1959.

1. Systems and vehicle familiarization — The seven astronauts were given lectures in the vehicle systems by NASA and several of the contracting companies. NASA Langley Research Center gave them a 50-hour course in Astronautics. McDonnell Aircraft Corp. engineers talked to the astronauts on Mercury subsystems. Lectures were given to the astronauts by Dr. William K. Douglas on aeromedical problems of space flight. At the Navy centrifuge in Johnsville, Pa. the astronauts flew the Mercury acceleration profiles. At several Air Force bases they flew brief zero-gravity flight paths. Checkouts of the Mercury environmental system and the pressure suit were accomplished at the Navy Air Crew Equipment Laboratory in Philadelphia. At the Naval Medical Research Institute they became familiar with the physiological effects of high CO_2 content in the environment. The Army Ballistic Missile Division and its associated contractors indoctrinated them on the Redstone. The Air Force Space Systems Division and its associated contractors told the astronauts about the Atlas launch vehicle.

2. Star recognition — Each astronaut periodically received concentrated personal instruction on the elements of celestial navigation and on star recognition at the Morehead Planetarium, Chapel Hill, North Carolina. A trainer simulating the celestial view through a spacecraft window permitted astronaut practice in correcting yaw drift.

3. Desert Survival — A 5½ day course in desert survival training was carried out at the USAF Training Command Survival

School at Stead Air Force Base, Nevada. The course consisted of survival techniques through lectures, demonstrations, and application in a representative desert environment. The Mercury survival kit was also evaluated during this period.

4. Egress training — During March and April, 1960 open water normal egress training was conducted in the Gulf of Mexico off Pensacola, Florida. Each astronaut made at least two egresses through the upper hatch (up to 10-foot swells were experienced). Water survival training was also accomplished in August, 1960 and December, 1961 at Langley. Each of the astronauts made underwater egresses, some of which were made in the Mercury pressure suit.

5. Specialty assignments — The astronauts contributed to the Mercury development program by working directly with Manned Spacecraft Center engineers and by attending NASA McDonnell coordination meetings and booster panel meetings in their specialty areas. Astronaut specialty areas are:

Carpenter — Communications equipment and procedures, periscope operation, navigational aids and procedures.

Cooper — Redstone booster, trajectory aerodynamics, countdown, and flight procedures, emergency egress and rescue.

Glenn — Cockpit layout, instrumentation, controls for spacecraft and simulation.

Grissom — Reaction control system, hand controller, autopilot and horizon scanners.

Schirra — Environmental control systems, pilot support and restraint, pressure suit, aeromedical monitoring.

Shepard — Recovery systems, parachutes, recovery aids, recovery procedures and range network.

Slayton — Atlas booster and escape system including Atlas configuration, trajectory, aerodynamics, countdown, and flight procedures.

MERCURY LAUNCH CHRONOLOGY

Two types of Mercury spacecraft have been used in the flight test program. First series of shots used full-scale "boilerplate" models of the capsule to check out booster spacecraft integration and the escape system. Second phase of the development firing program used Mercury capsules built to production standards.

This is the chronology of test firings:

September 9, 1959: Big Joe. NASA-produced research and development capsule, launched on an Atlas from Cape Canaveral — test validation of the Mercury concept. Capsule, survived high heat and airload and was successfully recovered.

October 4, 1959: Little Joe 1. Fired at NASA's Wallops Station, Virginia, to check matching of booster and spacecraft. Eight solid propellant rockets producing 250,000 lbs. of thrust drove the vehicle.

November 4, 1959: Little Joe 2. Also fired from Wallops Station, was an evaluation of the low-altitude abort conditions.

December 4, 1959: Little Joe 3. Fired at Wallops Station to check high-altitude performance of the escape system. Rhesus monkey Sam was used as test subject.

January 21, 1960: Little Joe 4. Fired at Wallops Station to evaluate the escape system under high airloads, using Rhesus monkey Miss Sam as a test subject.

May 9, 1960: Beach Abort Test. McDonnell's first production capsule and its escape rocket system were fired in an off-the-pad abort escape rocket system. (Capsule 1).

July 29, 1960: Mercury-Atlas 1. This was the first Atlas-boosted flight, and was aimed at qualifying the capsule under maximum airloads and afterbody heating rate during reentry conditions. The capsule contained no escape systems and no test subject. Shot was unsuccessful because of booster system malfunction (Capsule 4)

November 8, 1960: Little Joe 5. This was another in the Little Joe series from Wallops Station. Purpose of the shot was to check the production capsule in an abort simulating the most severe Little Joe booster and the shot was unsuccessful. (Capsule 3)

November 21, 1960: Mercury-Redstone 1. This was the first unmanned Redstone-boosted flight, but premature engine cutoff activated the emergency escape system when the booster was only about one inch off the pad. The booster settled back on the pad and was damaged slightly. The capsule was recovered for re-use. (Capsule 2)

December 19, 1960: Mercury-Redstone 1A. This shot was a repeat of the November 21 attempt and was completely successful. Capsule reached a peak altitude of 135 statute miles, covered a horizontal distance of 236 statute miles and was recovered successfully. (Capsule 2)

January 31, 1961: Mercury-Redstone 2. This was the Mercury-Redstone shot which carried Ham, the 37-lb, chimpanzee. The capsule reached 155 statute miles altitude, landed 420 statute miles downrange, and was recovered. During the landing phase, the parachuting capsule was drifting as it struck the water. Impact of the angle blow slammed the suspended heat shield against a bundle of potted wires, which drove a bolt through the pressure bulkhead, causing the capsule to leak. Ham was rescued before the capsule had taken on too much water. (Capsule 5)

February 21, 1961: Mercury-Atlas 2. This Atlas-boosted capsule shot was to check maximum heating and its effect during the worst reentry design conditions. Peak altitude was 108 statute miles; reentry angle was higher than planned and the heating was correspondingly worse than anticipated. It landed 1425 statute miles downrange. Maximum speed was about 13,000 mph. Shot was successful. (Capsule 6)

March 18, 1961: Little Joe 5A. This was a repeat of the unsuccessful Little Joe 5; it was fired at Wallops Station and was only marginally successful (Capsule 14)

April 25, 1961: Mercury-Atlas 3. This was an Atlas boosted shot attempting to orbit the capsule with a "mechanical astronaut" aboard. But 40 sec. after launching the booster was destroyed by radio command given by the range safety officer. The capsule was recovered and will be fired again(Capsule 8)

April 28 1961: Little Joe 5B. This was the third attempt to check the escape system under worst conditions, using a Little Joe booster fired from Wallops Station. Capsule reached 40,000 ft., and this time the shot was a complete success (Capsule 14)

May 5, 1961: Mercury-Redstone 3. This Redstone-boosted shot carried Astronaut Alan B. Shepard, Jr. on a ballistic flight path reaching a peak altitude of 116 statute mi. and a downrange distance of 302 statue mi. Flight was successful. (Capsule 7)

July 21, 1961: Mercury-Redstone 4. This successful flight carried Astronaut Virgil I. "Gus" Grissom to an altitude of 118 statute miles and 303 miles downrange. The capsule sank despite helicopter recovery efforts. (Capsule 11)

September 13, 1961: Mercury-Atlas 4. This successful flight saw the spacecraft attain orbit for the first time. The craft carried a "crewman simulator" designed to use oxygen and put moisture into the cabin at about the same rate as a man. Craft was recovered as planned about 160 miles east of Bermuda after one orbit. (Capsule 8)

November 29, 1961: Mercury-Atlas 5. The flight successfully carried the chimpanzee Enos through two orbits to a smooth landing. The craft was recovered about 260 miles south of Bermuda. (Capsule 9)

RESULTS OF THE FIRST UNITED STATES MANNED ORBITAL SPACE FLIGHT FEBRUARY 20,1962

Manned Spacecraft Center

NATIONAL AERONAUTICS AND SPACE ADMINISTRATION

FOREWORD

This document presents the results of the first United States manned orbital space flight conducted on February 20, 1962. The prelaunch activities, spacecraft description, flight operations, flight data, and postflight analyses presented form a continuation of the information previously published for the two United States manned suborbital space flights conducted on May 5, 1961, and July 21, 1961, respectively, by the National Aeronautics and Space Administration.

I. OPERATIONAL REQUIREMENTS AND PLANS

By JOHN D. HODGE, Flight Operations Division, NASA Manned Spacecraft Center; CHRISTOPHER C. KRAFT Jr., Chief, Flight Operations Division, NASA Manned Spacecraft Center; CHARLES W. MATHEWS, Chief, Spacecraft, Research Division, NASA Manned Spacecraft Center; and SIGURD A. SJOBERG, Flight Operations Division, NASA Manned Spacecraft Center

Summary

This paper presents a brief outline of the overall operational requirements and plans for the MA-6 mission. A short description of the tracking and ground instrumentation network is given together with general plans for recovery. Some aspects of flight-control training and simulation are discussed and requirements for weather forecasting and allowable weather conditions are examined. Total personnel commitments for direct operations are enumerated.

Introduction

The operational plan for Project Mercury has been presented in a number of papers previously (see refs. 1 and 2) and, therefore, it is only the intention of this paper to give a very brief outline of the preparations for the MA-6 flight. The operation was planned by a great number of different groups of people, both civilian and Department of Defense, and no effort will be made to describe the efforts of any individual group. The entire operation was a coordinated team effort and was only accomplished by the complete cooperation of all concerned.

Network Stations

In order to perform real-time analysis of both the powered phase and orbiting flight, a network of 16 stations located along the three orbit ground track was constructed. The location of these sites is shown in figure 1-1 and table 1-1. The sites were chosen for a great number of reasons but primarily to take advantage of the existing tracking and data-gathering facilities that were available at the start of the project. In addition, it was desirous to make available continuous real-time tracking during the launch and reentry phases, and to provide communications and telemetry data as often as possible throughout the flight. In addition, six of the sites at pertinent points along the ground track were provided with radio command capability in order to back up such functions as retrofire and clock changes. These were Atlantic Missile Range; Bermuda; Muchea, Australia; Hawaii; Guaymas, Mexico; and Southern California. A communications and computing center located at the NASA Goddard Space Flight Center, Greenbelt, Md., acts as the focal point for these network sites, and by means of telephone lines and microwave links provides high-speed information to the Mercury Control Center at Cape Canaveral, Fla. This network has proven to be an exceptionally good and capable tool in controlling orbiting vehicles and, in particular, manned spacecraft. (See appendix A.)

TABLE 1-1 — Ground Station for Project Mercury			
Stations	Voice and telemetry	S or C Band radar	Command capability
1. Atlantic Missile Range	X	C	X
2. Bermuda	X	SC	X
3. Mid-Atlantic Ship	X		
4. Canary Islands	X	S	
5. Kano, Nigeria	X		
6. Zanzibar	X		
7. Indian Ocean Ship	X		
8. Muchea, Australia	X	S	X
9. Woomera, Australia	X	C	
10. Canton Island	X		
11. Hawaii	X	S-C	X
12. Southern California	X	S-C	X
13. Guaymas, Mexico	X	S	X
14. White Sands		C	
15. Corpus Christi, Texas	X	S	
16. Eglin (AFATC)		C	

FIGURE 1-1.—Network stations distribution for Project Mercury

Recovery Forces

One of the biggest considerations for Mercury operations planning was to provide for safe and quick recovery of the astronaut. Plans were made to place forces in a large number of strategic locations to cover possible aborted flights during all phases of the mission. Figure 1-2 gives a general picture of the recovery operation. Plans were made for recovery in the launch area, on the basis of any foreseeable catastrophe from an off-the-pad abort to aborts occurring at or shortly after lift-off. From this period to the point at which the spacecraft was inserted into orbit, a number of recovery areas were located across the entire Atlantic Ocean, based on the probabilities of aborted flights occurring as a result of launch vehicle or spacecraft malfunctions. It was possible to choose discrete areas on the basis of using the spacecraft retrorockets to control the total range traveled. In addition, certain areas around the world were set up for contingencies should it be necessary to reenter the spacecraft as a result of some in-flight emergency. These areas are as indicated in figure 1-2.

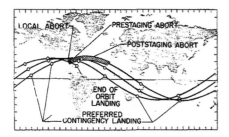

FIGURE 1-2.—Project Mercury recovery areas.

It was desirable to have the capability of recovery at the end of each of the three orbits, and primary landing areas located as shown were established for this purpose. The forces necessary to support the various recovery areas were based on the probability of having to end the flight in a particular phase of the mission, and the requirements for the various areas were normally given in terms of access time to the spacecraft once the landing had taken place. For instance, in the high probability area at the end of three orbits, a maximum of 3 hours was stated as the time to reach the spacecraft. On the other hand, the contingency area located on the east coast of Africa did not have pre-deployed forces, but airplane search capability was provided so that the spacecraft could be located in a maximum of 18 hours. In this case, the capability of providing emergency aid by means of paramedics was available should it be necessary. A more detailed description of the recovery operation actually performed in MA-6 is presented in paper 7.

Training

In the 3 years previous to the MA-6 flight, all of the operating elements supplying support to the operation have gone through various training exercises in preparation for a particular flight. Of course, the Mercury-Redstone flights, both unmanned and manned, and the Mercury-Atlas orbital flights previous to MA-6 were accomplished in preparation for the first manned orbital flight and provided the best and most realistic training. Also, both the astronaut and the ground flight-control personnel performed a great number of simulated launches and orbital flights in preparation for each of these exercises and were a highly trained and professional organization by the time the MA-6 flight was made.

Weather Information

As the whole world now knows, one of the most difficult operating problems encountered was the weather. Early in the project NASA solicited the aid of the U.S. Weather Bureau in setting up an organization to supply pertinent weather information. This group developed means for obtaining fairly detailed weather data along the entire three-orbit track of the Mercury mission. This information was analyzed in many different ways to provide useful operational information. For instance, detailed analysis of the weather over the Atlantic Ocean for various periods of the year was made to provide a basis of planning the flight and to provide a background knowledge as to what could be expected to develop from day to day once a given weather pattern had been determined. As a guideline, weather ground rules were established on the basis of spacecraft structural limitations and recovery operating capabilities. These included such details as wind velocity, wave height, cloud cover, and visibility. During the days previous to and on the day of the operation, the U.S. Weather Bureau meteorologists provided weather information for all of the preselected recovery areas and the launching site. The other weather limitation was the result of the desire to obtain engineering photographic coverage in the launch area.

Optical Tracking

Optical tracking of the Mercury-Atlas launch vehicle is part of the total launch instrumentation. These camera observations are used in conjunction with data from other instrumentation for establishing launch records.

TABLE I-II. Direct Operations Support Personnel

Agency	Location	Function	Number of personnel
National Aeronautics and Space Administration.	Canaveral	Launch support	300
	Worldwide	Network	50
	Worldwide	Flight control	90
	Worldwide	Recovery	15
Department of Defense	Canaveral	Launch support	1,900
	Worldwide	Network	400
	Worldwide	Recovery	15,600
	Worldwide	Aeromedical	160
Contractor	Canaveral	Launch support	380
	Worldwide	Network	360
Weapons Research Establishment	Australia	Network	50
Department of Defense utilizes considerable contractual support in these areas.			

In some cases where serious malfunctions occur, the photographic data are the only source for establishing the exact sequence of events.

Cameras are used to obtain trajectory parameters in the early phase of launch, for engineering sequentials and for historical documentation. They are located in the vicinity of the launch pad, in the general area of Cape Canaveral and along the Florida Coast both North and South of the launch area.

Support Personnel

As an indication of the total effort involved in the Mercury operation, it is interesting to note the number of people who participated. A total of about 19,300 people were deployed at the time of the mission. By far the largest number (about 15,600) were associated with the recovery effort. About 2,600 were involved at the launch complex and 1,100 were manning the tracking network. (See table I-II)

Details of spacecraft and launch-vehicle preparations and the flight plan and control of the flight are described in subsequent papers.

2. SPACECRAFT AND SPACECRAFT SYSTEMS

By KENNETH S. KLEINKNECHT, Manager, Mercury Project, NASA Manned Spacecraft Center; WILLIAM M. BLAND, Jr., Deputy Manager, Mercury Project, NASA Manned Spacecraft Center; and E. M. FIELDS, Chief, Project Engineering Office, Mercury Project, NASA Manned Spacecraft Center

Summary

The Mercury spacecraft used by Astronaut John H. Glenn, Jr., in successfully accomplishing the first manned orbital flight from the United States performed as it was designed. Performance of all systems was at least as good as design, and in some cases better, as for instance, communications and manual attitude control. Deleterious effects of minor systems malfunctions were effectively avoided by system redundancy aided by astronaut corrective action, as in the case of the malfunctioning control system, and by ample system design margins, as in the cases of the lack of inverter cooling and non-design reentry without jettisoning the retropackage.

Introduction

The Mercury spacecraft is designed to sustain a man in a space environment for a given period of time, to protect him from external heating and acceleration during exit and reentry, to provide him with means for controlling the attitude of the spacecraft, to permit him to perform observations and a limited number of experiments in space, and to then bring him safely back to earth with sufficient location aids to permit rapid recovery by surface forces. The purpose of this paper is to present a brief description of the spacecraft and its systems and to provide a limited description of the performance of the spacecraft during the first manned orbital flight from the United States.

FIGURE 2-1 — Exterior of spacecraft 13.

The external arrangement of the spacecraft is shown in figure 2-1. At the time this photograph was taken the spacecraft was mounted on the launch vehicle and was undergoing final preparation for flight. The spacecraft is just large enough to contain the astronaut and the necessary equipment. The main conical portion contains the crew, the life-support system, the electrical-power system, and necessary displays and system controls. The cylindrical section contains the major components of the parachute landing system. The topmost section contains a bicone antenna for RF transmission and reception and the drogue (stabilizing) parachute which is deployed during the landing phase.

The large face of the spacecraft is protected against reentry heating by an ablation-type heat shield. A package which contains three posigrade rocket motors and three retrograde rocket motors is held to the spacecraft at the center of the heat shield by three straps.

The spacecraft escape system includes an escape rocket motor on top of a tower which is fastened to the top of the recovery section by a clamp ring. The escape-tower assembly also incorporates a small rocket motor which jettisons the tower should the escape motor be fired. In a normal mission such as MA-6, the escape tower is jettisoned by firing the escape rocket motor soon after

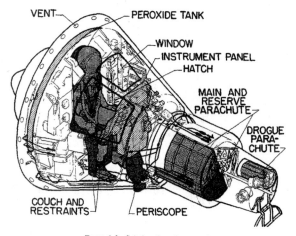

FIGURE 2-2.—Interior view of spacecraft.

launch vehicle staging when the aerodynamic forces have decreased so that the escape rocket motor is no longer required for a possible abort maneuver. The spacecraft posigrade rocket motors provide for separation from the launch vehicle after this time for either aborted or normal missions.

The entire spacecraft is mounted to a special adapter section on the Atlas launch vehicle and is restrained by an explosively actuated clamp ring.

Figure 2-2 shows an interior view of the spacecraft. The astronaut is supported by a molded couch and other restraints and faces the small end of the spacecraft. Accelerations during both exit and reentry act in the same direction and thus enable his support couch to be effective without reorientation. The astronaut faces a display of the surface of the earth through a periscope and an instrument panel as shown in figure 2-3.

The instrument panel, which is shown in figure 2-4, is supported by the periscope structure. It contains the instruments and display lights necessary to monitor spacecraft systems and sequencing, the controls required to initiate primary sequences manually, and the necessary flight control displays.

Within the pressurized compartment, the major systems near the astronaut are batteries for d-c electrical power, the environmental control system, and major components of the communications and instrumentation systems. The astronaut operates control sticks with each hand. The right-hand stick is used for manually controlling the spacecraft attitude, and the left-hand stick can be used for initiating the escape sequence in event of an emergency.

Between the pressure compartment and the heat shield are the tanks for the hydrogen peroxide which is used as fuel for the attitude control system. In addition, the landing bag is folded up and stowed in this area. Around the periphery of the large pressure bulkhead are vents for the steam which is given off by the environmental control system.

A window is provided, in the conical section over the astronaut's head, for the astronaut to use for observations and for obtaining visual attitude references. The astronaut entrance and egress hatch is located in the conical section as indicated by the outline to the astronaut's right. The astronaut can also egress through the recovery compartment by removing a portion of the instrument panel, the forward pressure bulkhead, and the parachute container.

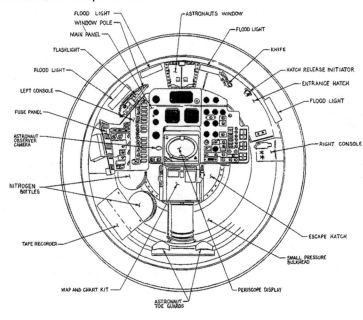

FIGURE 2-3.—Spacecraft cabin arrangement.

The spacecraft at the launch of MA-6 weighed 4,265 pounds. The weight at insertion into orbit was 2,987 pounds, and at retrograde, 2,970 pounds. The water landing weight was 2,493 pounds, and the recovery weight was 2,422 pounds.

HEAT PROTECTION

An artist's conception of the Mercury spacecraft during the early stages of a normal reentry is shown in figure 2-5, with shading indicating that the spacecraft is surrounded by a bright-orange envelope of heated air. The spacecraft has been designed to protect the interior from the effects of reentry aerodynamic heating. This heat protection consists of an ablation reentry heat shield for the forebody and an insulated double-wall structure for the afterbody.

Ablation Shield

The ablation-shield material is a mixture of glass fibers and resin in the proper proportions such that the resin will boil off under applied heat with the glass fibers to provide strength and shield integrity. During the high heating period of reentry, the resin vaporizes and boils off at low temperatures into the hot boundary layer of air thus cooling.

The shield is designed to withstand the heat loads generated by more severe reentry conditions than those experienced during the MA-6 mission when only a few pounds of the shield were boiled away

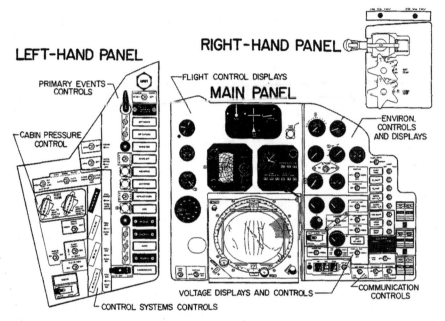

FIGURE 2-4.—Instrument panels.

Afterbody

FIGURE 2-5.—Artist's conception of spacecraft during reentry.

The afterbody (cone, cylinder, and antenna canister) is protected somewhat from the hot boundary layer of air and gaseous ablation products during reentry since most of the afterbody, with its inward sloping sides, is in the dead-air region behind the ablation shield. The afterbody surface thus receives only 5 to 10 percent of the heating that is experienced by the ablation shield; therefore its heat protection arrangement can be the more conventional double-wall construction with insulation between the inner and outer walls (see fig. 2-6). The Mercury spacecraft afterbody heat protection, as shown in figure 2-6, consists of a double-wall construction with insulation between the outer and inner walls. In figure 2-6 lightweight fiberglass blankets are labeled "insulation," and compressed surface-clad insulation is labeled "Min-K." On the outer conical surface and antenna section thin high-temperature alloy (René 41) shingles are used. On the outer cylindrical section thicker shingles of beryllium are used in a heat-sink arrangement. The shingles are blackened to aid the radiation of heat away from the spacecraft, and they are attached to the basic structure in such a manner that they can expand and contract with temperature changes without transferring loads to the primary load carrying spacecraft structure.

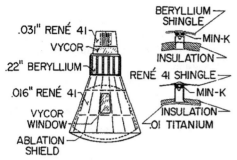

FIGURE 2–6.—Arrangement of heat-protection elements.

In order to illustrate the effectiveness of the spacecraft heat protection, some significant temperatures are presented in figure 2-7. From an overall standpoint, the most severe beating is encountered during reentry. During this time, the air cap surrounding the front end of the spacecraft has a maximum temperature of about 9,500° F, which is nearly the same as the temperature of the surface of the sun. As a result, the surface of the heat shield reaches a maximum temperature of about 3,000° F and the spacecraft afterbody shingles attain maximum temperatures in the order of 1,000° F on the thin shingles and about 600° F on the thicker shingles.

During the exit flight, when the small end of the spacecraft points in the direction of flight, the afterbody shingles are also subjected to aerodynamic heating, attaining maximum temperatures as high as about 1,300° F. With the local temperatures dependent upon local flow conditions and the thermal mass of the spacecraft surface.

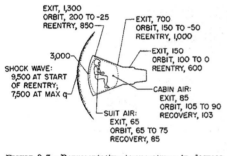

FIGURE 2–7.—Representative temperatures in degrees Fahrenheit.

The temperature variation of the outer shingles around the astronauts pressure compartment is modest during the orbital phase of the mission, varying between 200° F and -50° F, depending upon the sun impingement. Of particular interest are the cabin-air and suit-air temperatures. These remain at acceptable levels during all phases of the mission, attesting to the effectiveness of the environmental control system and the insulation.

Figure 2-8 shows the maximum temperatures in the ablation shield near the stagnation point during and after reentry for three orbital re-entries. As can be seen, preflight, calculations predicted quite well the measured temperatures. As is now well known, the MA-6 reentry was intentionally begun with the retropack in place in the center of the shield. Thus, it is particularly interesting to note that the maximum temperature measured near the stagnation point in the MA-6 shield was not much lower than those measured during the MA-4 and MA-5 re-entries. This and other available evidence indicates that the retropack disintegrated from reentry heating during the early part of the reentry so that its presence made little difference in the total heat input to the shield.

SPACECRAFT SYSTEMS

Rocket Motor Systems

The rocket motor assemblies used in the Mercury spacecraft are as shown in figure 2-9 and are, listed in the following table along with their nominal performance characteristics.

Rocket motor	Number of motors	Nominal thrust each, lb	Approximate burning time each, sec
Escape	1	52, 000	1
Tower jettison	1	800	1.5
Posigrade	3	400	1
Retrograde	.3	1,000	10

All of these rocket motors employ solid propellant fuel.

The escape rocket is mounted at the top of the escape tower and incorporates three canted exit nozzles to direct the exhaust gases away from the side of the spacecraft.

The tower jettison rocket also has a three nozzle assembly and is attached to the bottom of the escape rocket-motor case.

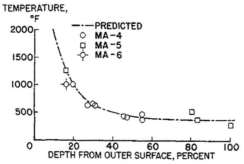

FIGURE 2-8.—Maximum ablation-shield temperatures.

The three posigrade and three retrograde rocket motors are mounted in a package which is held to the spacecraft at the center of the heat shield by three straps. The posigrade rockets are salvo-fired to separate the spacecraft from the launch vehicle. The retrograde rocket motors are ripple-fired (5-second delays between motor ignitions) and provide the velocity decrement necessary to initiate reentry.

All rocket motors have dual ignition systems from separate electrical power sources. In addition, each ignition system has dual squibs to insure ignition. A typical example of rocket motor system firing circuitry is shown in figure 2-10.

In the MA-6 mission, all rocket motor systems appear to have operated properly. The escape tower was jettisoned as planned at the proper time. The firing of the posigrade rockets provided the expected velocity change, as did the retrorocket motors.

Control System

The control system of the Mercury spacecraft provides the capability of performing several functions vital to a successful orbital mission; they are attaining a precise

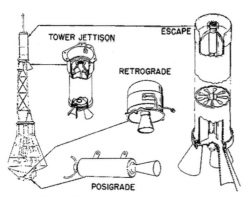

FIGURE 2-9.—Spacecraft rocket motors.

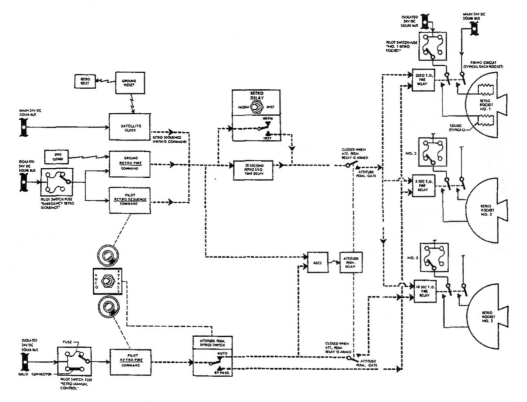

FIGURE 2-10.—Retrorocket fire schematic diagram.

attitude for retrofire and holding the attitude closely during the stepped thrusting period of the retrorockets. Without such control, an orbital mission would very probably suffer mission failure. Because of this critical function, the Mercury control system has been designed so that it can perform its function in event of multiple malfunctions.

Table 2-1 indicates the four control arrangements that are available in the present Mercury spacecraft. Basically, there are two completely independent fuel, supply, plumbing, and thruster systems. Each uses 90-percent hydrogen peroxide to provide selected impulse as desired. There are two means of controlling the outputs of each of these systems; that is, on system A the astronaut has a choice of using either the automatic stabilization and control system (ASCS) or the fly-by-wire (FBW) system. The ASCS is automatic to the extent that it can provide the necessary attitude control throughout a complete mission without any action on the part of the astronaut; this is the system that was used on the unmanned Mercury missions. The FBW system is operated by movement of the astronaut control stick to operate the solenoid control valves electrically.

TABLE 2-1 — Spacecraft Control System Redundancy and Electrical Power Requirements		
Control system modes	Corresponding fuel System (fuel supply, plumbing, and thrusters)	Electrical Power Required
ASCS ------	A ----------------	d-c and a-c
FBW ------	A ----------------	d-c
MP -------	B ----------------	None
RSCS ------	B ----------------	d-c and a-c
ASCS - Automatic stabilization and control system FBW- Fly-by-wire MP- Manual Proportional system - Controlled by pilot actuation of control RSCS - Rate stabilization control system stick		

On system B the astronaut has the choice of using either the manual proportional system (MP) or the rate stabilization control system (RSCS), both of which are operated through the astronaut's control stick. In the MP system, linkages transmit the control stick movement to proportional control valves which regulate the flow of fuel to the thrusters. The RSCS uses a combination of stick positions and the computing components of the automatic system to provide rate control.

The mode of control can be easily selected by the astronaut by positioning of the proper switches and valves mounted on the instrument panel. It should also be noted that certain of these control modes can be selected to operate simultaneously, such as ASCS and MP, or FBW and MP, in order to provide double authority or so that even with certain malfunctions in each mode, complete control can be maintained. Also of interest is the type of electrical power requirement for each of these control arrangements. Most significant is the lack of any electrical power requirement for the manual proportional control mode.

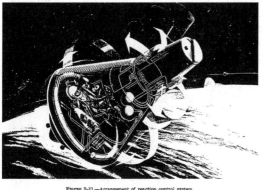

Figure 2-11.—Arrangement of reaction control system.

The thruster impulse is directed by the four basic control modes through 18 individual system, as shown in figure 2-11. Figures thrusters — 12 on system A and 6 on the manual 2-12(a) and 2-12 (b) show the A and B systems RCS schematic diagrams. Figures 2-12 (a) and 2-12(b) do not show completely the methods of electrically controlling the thrusters; in figure 2-12(a) there is a switch missing between the hand controller and the ASCS control box which permits selection of either FBW or ASCS. Similarly, in figure 2-12(b) the connection of the hand controller through a switch and the ASCS control box to the thruster solenoids for RSCS control is missing. Metered quantities of hydrogen peroxide are decomposed in silver-plated catalyst beds in each of the thruster chambers to provide the desired impulse. Twelve of the thrusters used on the Mercury spacecraft are sized to provide adequate control during the retro maneuver.

These thruster ratings are as follows:

Axis	System A, lb	System B, lb
Pitch	24	4 to 24
Yaw	24	4 to 24
Roll	6	1 to 6

The remaining six thrusters are in system A to provide fine attitude control as desired under orbital conditions with minimum fuel consumption. Each of these six thrusters has a thrust rating of 1 pound.

On the MA-6 mission the control system, with essential mode changes by the astronaut, provided adequate control of spacecraft attitudes during all phases of the mission despite recurrences of small thruster malfunctions which disabled the minimum fuel consumption mode about the yaw axis early in the mission. As discussed in paper 11, the astronaut very satisfactorily completed his planned maneuvers in space, orientated the spacecraft as he desired to make terrestrial and celestial observations, attained and maintained accurate control for retrofire by using both automatic and manual attitude-control modes, and accurately achieved entry attitude. During entry, after maximum dynamic pressure, the astronaut while on manual proportional and fly-by-wire control successfully controlled the lateral oscillations until the B-system fuel supply was depleted. At this time the oscillations began to build up; however, switching to an automatic mode did reduce oscillations to within desirable limits until the A-system fuel supply was also depleted.

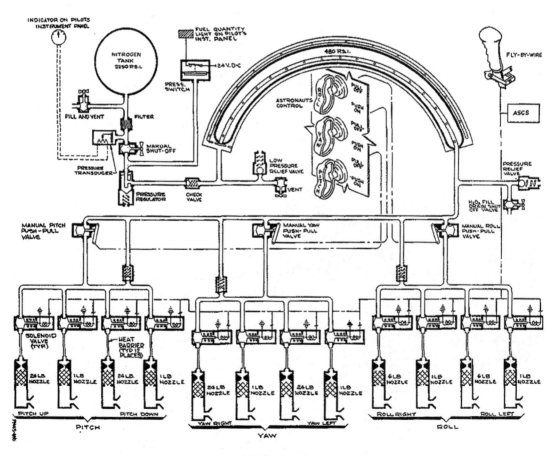

(a) System A.

FIGURE 2-12.—Reaction control system schematic diagrams.

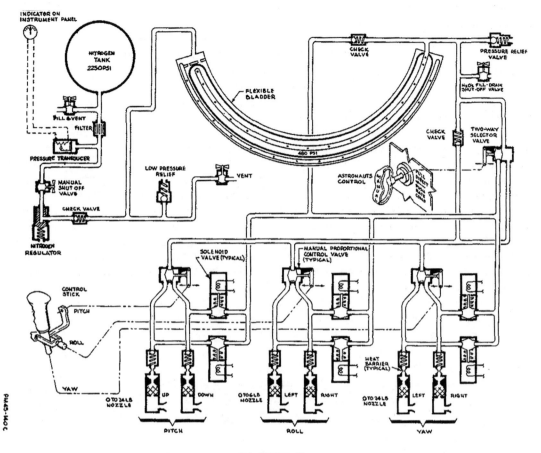

(b) System B.

Figure 2-12.—Concluded.

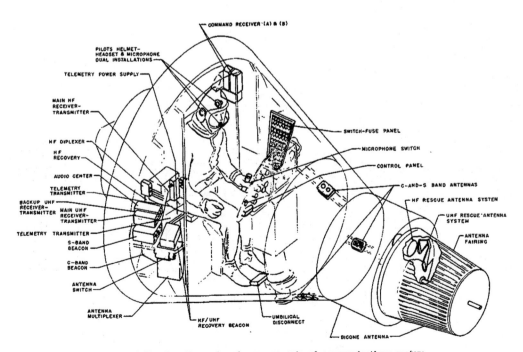

Figure 2-13.—Locations of major components of communications system.

Communications and Instrumentation

Communications

The spacecraft communications and instrumentation systems consisted of voice, radar, command, recovery, and telemetry links. Each system had main and backup (or parallel) equipment for redundancy, with selection of the desired system generally at discretion of the astronaut through switches mounted on the instrument panel. Table 2-11 is a list of the systems, and figure 2-13 shows the physical location of the communications equipment in the spacecraft. The performance of the various communication links was generally very good during the MA-6 mission as shown in figure 2-14. The time for which useful signals were obtained for each link are compared with the time that the spacecraft was within line-of-sight for each pass over two selected range stations. This same comparison was made, for each pass over each range station, averaged for the entire mission, and presented in percent form in the right hand column of figure 2-14.

TABLE 2-II. — Spacecraft Communications and Instrumentation Systems

Voice communication	
UHF transceiver-	2 watts
UHF transceiver-	0.5 watts
HF transceiver	5 watts
Radar	
C-band beacon	400-watt transponder
S-band beacon	1,000-watt transponder
Commands	
2 command receivers	10 channels each
Telemetry	
2 FM transmitters	2 watts each
Recovery	
HF D/F beacon SEASAVE	1 watt
UHF D/F beacon SARAH	7.5 watts
UHF D/F beacon SUPER SARAH.	91 watts
HF transceiver	1 watt

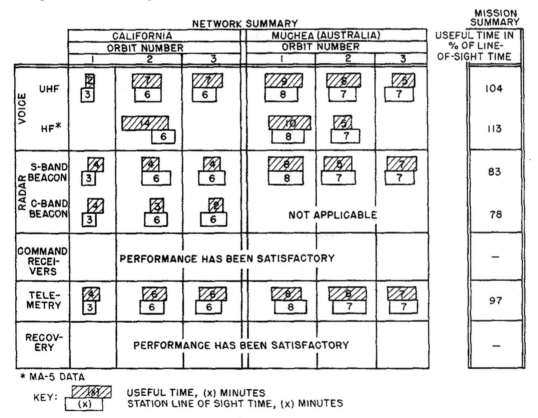

FIGURE 2-14.—Performance of spacecraft communications links.

Since the HF voice system was not used enough in the MA4 flight to allow a meaningful assessment of HF coverage, the HF coverage shown in figure 2-14 is for the MA-6 orbital flight which utilized a tape recording for spacecraft voice broadcasts on both HF and UHF systems. A simplified schematic diagram showing the various communications systems and their respective antenna systems is shown in figure 2-15.

Voice system. — The voice system, used for two-way voice conversations between the ground and spacecraft, was made up of high frequency (HF) and ultra-high-frequency (UHF) systems. From previous orbital experience and ground tests, it was known that the HF system had somewhat poorer voice fidelity but longer range than the UHF system. The UHF system, because of its slightly better voice quality, was considered to be the primary system. From previous experience, it was known that the range of the UHF system was approximately equal to the line-of-sight range and was entirely adequate for a normal mission. The main voice traffic was therefore conducted on the UHF system, with a small amount of traffic conducted on the HF system to verify system operation. It might be noted that the UHF system consisted of a main and a backup transmitter receiver. Thus, three separate voice systems were available for choice of the astronaut — HF, UHF main, and UHF backup-as shown in figure 2-16. Also, a redundant ground-to-air voice link is available through the command receiver channel. An additional air-to-ground communication link is available to the astronaut by keying the low-frequency telemeter carrier. It should be noted that all of these links use the main bicone antenna through the use of a multiplexer.

Performance of the voice system during the MA-6 mission was satisfactory as indicated in figure 2-14.

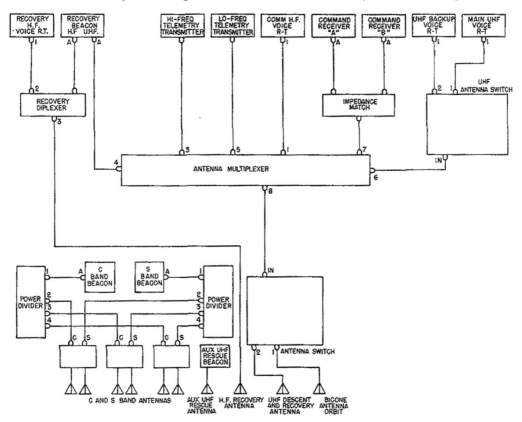

FIGURE 2-15.—Communications and antenna systems schematic diagram.

Radar system. — The radar system consisted of C and S band beacons onboard the spacecraft. Both beacons were "on" continuously throughout the flight, and either or both beacons could be interrogated when within range of the appropriate ground station. Performance of the radar system, including the ground tracking and computing complex, was satisfactory and was such that the spacecraft orbital trajectory was well defined by the end of the first orbit, and continued tracking during the remaining orbits resulted in only minor changes to orbit parameters already established. Radar performance is shown in figure 2-14 and is discussed in more detail in paper 6.

Command system. — The command system provided means of commanding an abort, retrofire, spacecraft - clock change, or instrumentation calibration from the ground, if necessary. None of the first three of the above commands were needed for accomplishment of the MA-6 mission. The onboard command system was exercised by ten instrumentation-calibration ground commands during the mission for instrument calibration and to obtain additional data on the command-system in-flight performance. The onboard command system consisted of two identical receivers and decoders, each capable of performing any required functions. The data are being studied to evaluate the command-system performance, which has been satisfactory in previous orbital missions.

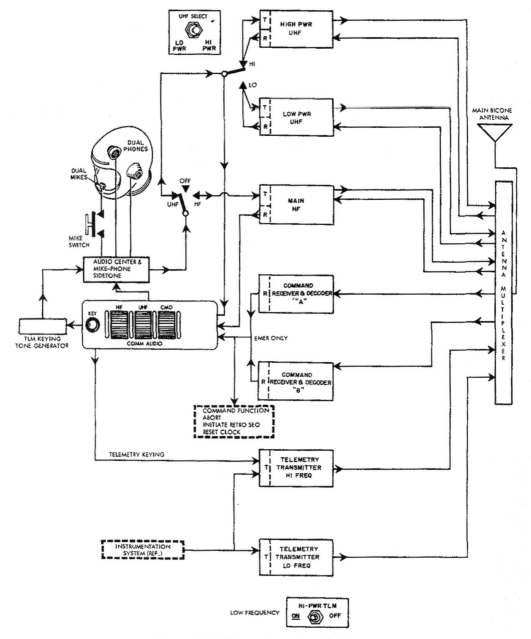

FIGURE 2–16.—Voice systems schematic diagram.

Telemetry system. — The telemetry system consisted of two nearly identical FM-FM telemetry subsystems, each carrying essentially the same data for redundancy and each using commutated and continuous channels. The aeromedical data and some system performance information from the telemetry system were displayed in real time at the Mercury Network stations for the purpose of monitoring the conditions of the astronaut and critical spacecraft systems as the mission progressed. The performance of the telemetry system was satisfactory as shown in figure 2-14.

Recovery system. — The recovery system consists of a HF transceiver (1 watt), one recovery package containing the CW SEASAVE beacon (1 watt), a pulse modulated SARAH beacon (7.5 watts), and a pulse modulated SUPERSARAH beacon (91 watts). The antenna systems used by the recovery system are shown in figure 2-15. Performance of the recovery system has been satisfactory.

Instrumentation

The spacecraft instrumentation system monitored the astronaut's ECG, respiration rate and depth, blood pressure, and body temperature in addition to certain aspects of operations of the spacecraft systems. Locations of many of the sensors are shown in figure 2-17, and the instrumentation list is shown in table 2-III. Ninety commutator segments were available for data, plus seven continuous channels. The continuous channels were used mainly for aeromedical information and spacecraft control-system performance data. An instrumentation schematic diagram is included as figure 2-18, to show how the different measurements were handled. A 16-mm camera photographed the astronaut's face and upper torso area in color at 360 frames/min or 5 frames/min, depending on the mission phase. The overall quality of these photographs was good; however, due to extreme variations in the light intensities in the spacecraft during the various mission phases, definition was reduced at times. Performance of the instrumentation system was satisfactory.

TABLE 2-III- *Spacecraft Instrumentation and Ranges for the MA-6 Mission*
[All data commutated, unless otherwise noted. Instrument range (0 to 100 percent full scale unless otherwise noted)

Events:

Tower release	On-off
Tower escape rocket ignition	On-off
Spacecraft separation	On-off
Retro-attitude command	On-off
Retrorocket fin	On-off
Retrorocket assembly jettison	On-off
0.05g relay	On-off
Drogue parachute deployment	On-off
Antenna fairing release	On-off
Main parachute deployment	On-off
Periscope retract	On-off
Mayday	On-off
Heat-shield deployment	On-off
Main parachute jettison	On-off
Reserve parachute deploy	On-off
Pilot abort	On-off
Stabilization, and control:Control stick position (roll)1, deg	±12
Control stick position (pitch)1, deg	±12
Control stick Position (yaw)1, deg	±14
Gyro output (roll), deg	- 130 to 190
Gyro output (pitch), deg	- 120 to 174
Gyro output (yaw), deg	- 70 to 250
Scanner output (roll), deg	- 37.5 to 33
Scanner output (pitch), deg	- 38.5 to 33
Scanner ignore (roll)	On-off
Scanner ignore (pitch)	On-off

ASCS slaving signal	On-off
Roll solenoid, high	On-off
Roll solenoid, high	On-off
Pitch solenoid, high	On-off
Pitch solenoid, high	On-off
Yaw solenoid, high	On-off
Yaw solenoid, high	On-off
Roll solenoid, low	On-off
Roll solenoid, low	On-off
Pitch solenoid, low	On-off
Pitch solenoid, low	On-off
Yaw solenoid, low	On-off
Yaw solenoid, low	On-off
Roll rate (low range) 2, deg/sec	-9.9 to 10
Roll rate (high range)2, deg/sec	-25.5 to 31.5
Pitch rate 2, deg/sec	- 10.3 to 10.5
Yaw rate 2, deg/sec	- ±10.3

Electrical functions:

3 volt reference	100-percent full scale
Zero reference	0-percent full scale
7 volt a-c bus, volts	0 to 8
d-c 3 , volts	12.9 to -29.8
d-c current 3, amperes	0 to 50
Fans a-c bus, volts	95 to 120
ASCS a-c bus, volts	90 to 125
Isolated d-c bus, volts	13.3 to 23.4
Standby d-c bus, volts	13.2 to 23
Standby inverter ON	On-off

Astronaut:

Body temperature, °F	92.5 to 105.9
Body temperature, °F	92.6 to 106
ECG #1 4	—
Respiration 4	—
Blood Pressure 4, mm Hg	56 to 205

Command receivers:

Command receiver ON-OFF	On-off
Command receiver signal strength "A", muV	0 to 80
Command receiver signal strength "B", muV	0 to 80

Environmental functions:

Suit inlet temperature, °F	39 to 116
Cabin temperature, °F	35 to 233
Suit pressure, psia	0 to 15
Cabin pressure, psia	0 to 15
Static pressure, psia	15.2 to -0.3
Coolant quantity pressure psig	213 to 489
O2 supply pressure, primary, psig	-50 to 7,500
O2 supply pressure, secondary, psig	-100 to 7,600
O2 partial pressure, mm Hg	- 15 to 980
O2 emergency rate mode	On-off

Accelerations:

Acceleration, Ax (low range), g units	- 0.415 to 0.375
Acceleration, Ax (high range), g units	- 3.2 to 3.5
Acceleration, Ay (low range), g units	- 0.415 to 0.38
Acceleration, Ay (high range), g units	- 4.0 to 4.9
Acceleration, A	- 31 to 35
Integrating accelerometer signal ft/sec	0 to 565

Equipment temperatures:

RCS automatic H_2O_2 line temperature low roll, clockwise, °F	- 6 to 242
RCS automatic H_2O_2 line temperature low roll, counterclockwise, °F	- 11 to 239
RCS manual H_2O_2 line temperature low roll, clockwise, °F	- 12 to 260
RCS manual H_2O_2 line temperature low roll, counterclockwise, °F	-29 to 224
Retrorocket temperature, °F	- 16 to 140
Heat-shield temperature, °F	- 140 to 2,470
Inverter temperature, 150 v-amp, °F	- 10 to 337
Inverter temperature, 250 v-amp, °F	- 18 to 322
Transmitter temperature, HF, °F	8 to 320
Transmitter temperature, LF, °F	-16 to 326
Onboard time: Verner clock 4, percent	0 to 12
Time since launch, percent	0 to 100
Time to retrograde, percent	0 to 100

Calibrate signal

1 Commutated and continuous; continuous data recorded onboard only.

2 Continuous only; recorded onboard only.

3 Commutated and continuous; continuous d-c current and d-c voltage recorded onboard only.

4 Continuous only.

Environmental Control System

As shown in figure 2-19, the environmental control system is located principally under the astronaut's support couch. This all-important system provides an oxygen atmosphere, temperature control, and pressure regulation for the astronaut's suit and the cabin. The cabin and suit are independent redundant circuits for automatically providing proper environmental conditions for the astronaut. In addition, the astronaut can manually actuate a control to initiate the oxygen emergency-flow-rate mode to provide an adequate suit environment in the case where multiple plumbing or electrical failures might make automatic initiation of the emergency-flow-rate mode inoperative.

The environmental control system provided, with its automatic operation, an adequate and safe environment for the astronaut throughout the entire MA-6 mission.

A simplified schematic diagram of the environmental control system is shown in figure 2-20, and the system is discussed in more detail in papers 3 and 5.

Electrical Power and Sequential Systems

Electrical Power System

Figure 2-21 shows the spacecraft electrical power system. Rechargeable silver-zinc batteries of both 3,000 watt-hour and 1,500-watt-hour ratings are arranged into three power sources to provide a total of 13,500 watt-hours. Nominal discharge rate for each type of battery is 4.5 amperes but each battery is capable of supplying pulse currents up to 42 amperes for a few milliseconds. Silicon diodes in each positive leg of all batteries prevent discharge of normal batteries into defective or low voltage batteries in parallel configurations. The individual battery voltages and the total current are monitored in flight by the astronaut. In event of battery failure or equipment failure, he can manually switch off individual batteries or all battery power.

A solid-state static inverter provides 115 volt, 400-cycle, single-phase alternating-current, power for the spacecraft attitude control system and another provides similar power for the environmental control system.

A standby inverter provides redundancy for either of these two main inverters and can take over for both if non-critical ASCS loads are switched off manually. The standby inverter can be automatically or manually placed in action.

Electrical power consumption for the MA-6 mission is shown in the following table:

Battery system	Prelaunch		Orbital		Postlanding		Total		Reserve [1]	
	Watt-hours expended	Available power used, percent	Watt-hours expended	Available power used, percent	Watt-hours expended	Available power used, percent	Watt-hours expended	Available power used, percent	Watt-hours	Percent of total power
Main and standby (paralleled)	606	5	2,480	20.6	260	2.2	3,346	27.8	8,654	72.2
Isolated	30	2	50	3.3	40	2.7	120	8	1,380	92

[1] The reserve percentages reflect the relatively short prelaunch and recovery periods for MA–6. With allowances for a 2-hour prelaunch period and a 12-hour recovery period, battery reserves would become 50 percent for the main and standby supplies and 47 percent for the isolated supply, which are considered very adequate margins.

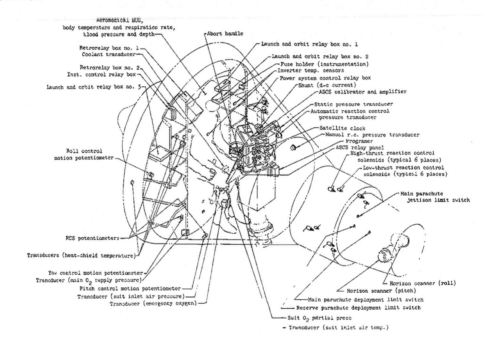

FIGURE 2–17.—Location of some spacecraft instrumentation sensors.

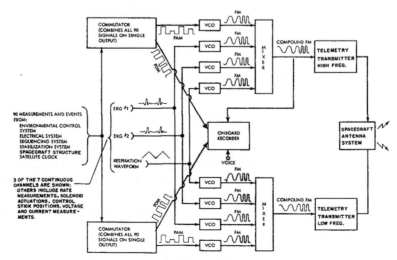

FIGURE 2–18.—Instrumentation system schematic diagram.

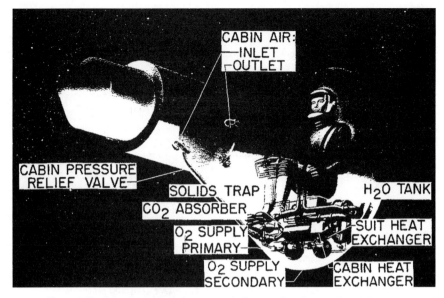

FIGURE 2-19.—Approximate physical arrangement of spacecraft environmental control system.

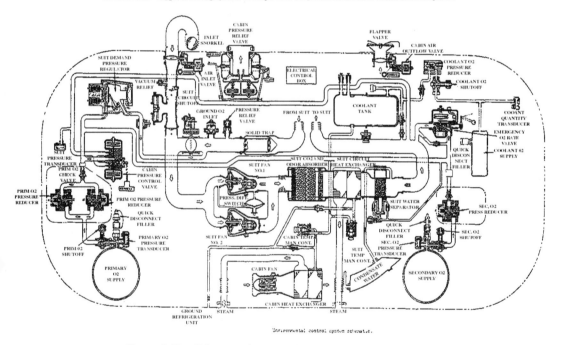

FIGURE 2-20.—Schematic diagram of environmental control system.

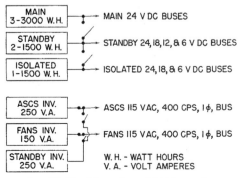

FIGURE 2-21.—Schematic diagram of spacecraft electrical power system.

Both the d-c and a-c electrical systems functioned well throughout the MA-6 mission. Excessive temperature rises of the inverters were caused by a malfunction of the inverter cooling system. The inverter design operating temperatures was exceeded for both inverters during the second orbit. Maximum inverter temperatures were over 200° F, somewhat higher than design temperature; however, performance of the inverters during the mission was excellent and no adverse effects due to the high temperatures were noted during postflight inspections.

Sequential System

Figure 2-22 shows the sequence of major events for the MA-6 mission. The only exception to the planned sequence occurred during event (9) as a result of the astronaut being advised to retain the retropackage during reentry and manually overrode the automatic sequencing as directed by instructions from the ground. Figure 2-23 shows the spacecraft master sequential system.

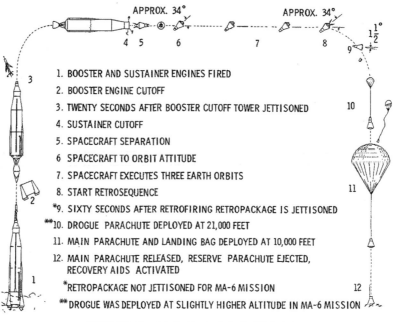

APPROX. 34°   APPROX. 34°

4 5   6   7   8   9

1. BOOSTER AND SUSTAINER ENGINES FIRED

2. BOOSTER ENGINE CUTOFF

3. TWENTY SECONDS AFTER BOOSTER CUTOFF TOWER JETTISONED

4. SUSTAINER CUTOFF

5. SPACECRAFT SEPARATION

6 SPACECRAFT TO ORBIT ATTITUDE

7. SPACECRAFT EXECUTES THREE EARTH ORBITS

8. START RETROSEQUENCE

*9. SIXTY SECONDS AFTER RETROFIRING RETROPACKAGE IS JETTISONED

**10. DROGUE PARACHUTE DEPLOYED AT 21,000 FEET

11. MAIN PARACHUTE AND LANDING BAG DEPLOYED AT 10,000 FEET

12. MAIN PARACHUTE RELEASED, RESERVE PARACHUTE EJECTED, RECOVERY AIDS ACTIVATED

*RETROPACKAGE NOT JETTISONED FOR MA-6 MISSION

**DROGUE WAS DEPLOYED AT SLIGHTLY HIGHER ALTITUDE IN MA-6 MISSION

FIGURE 2-22.—Sequence of major events for the MA-6 mission.

TABLE 2-IV. — Methods of Initiating or Controlling Major Mission Sequences

Major mission events	Initiation capability		
	Automatic	Ground	Pilot
1 Booster and sustainer engines firing	------	x	-----
2 Booster engine cutoff	xx	x	xxx
3 Tower jettison	x	------	x
4 Sustainer engine cutoff	xx	x	xxx
5 Spacecraft separation	x	------	x
6 Yaw and pitch maneuver to orbit attitude	x	------	x
7 Attitude control during orbit	x	------	x
8 Start retrosequence	x	x	x
9 Retropackage jettison and reentry attitude control	x	----	x
10 Drogue parachute deployment	x	------	x
11 Main parachute and landing bag deployment.	x	------	x
12 Main parachute release, reserve parachute ejection, recovery aids activation	x	------	x
xx Refers to automatic redundant system. xxx Refers to indirect control. x Refers to direct control.			

Redundancy for the automatic initiation of spacecraft sequence events is furnished by the astronaut's ability to initiate events manually by switches and controls and by ground-commanded initiation of certain events by means of a radio-command link. Table 2-IV shows the redundancy in initiation capability for important events. From this tabulation it can be seen that the astronaut has control over all of the primary spacecraft automatic functions. The astronaut also has indirect control over the important launch-vehicle functions, such as engine cutoff, through use of the spacecraft abort handle when necessary.

The performance of the spacecraft sequencing system was satisfactory during the MA-6 mission.

Landing System

The landing system consists of the drogue parachute, main and reserve parachutes, landing bag, and attendant functional systems. The landing system is armed when the escape tower is jettisoned during exit flight; however, it is not actuated until the spacecraft returns to the relatively dense parts of the earth's atmosphere, as shown in figure 2-24.

The landing system is normally actuated at an altitude of about 21,000 feet by either one of two barostats which sense atmospheric pressure. At this time, the drogue parachute is deployed to decelerate and stabilize the spacecraft. At about 10,000 feet, the antenna section and drogue parachute are jettisoned by the signals from another dual barostat and the main parachute is deployed in a reefed condition opened to 12-percent of the maximum diameter for 4 seconds to minimize the opening shock. The main parachute deploys fully after 4 seconds of reefing. A reserve parachute may be deployed by the astronaut in the event the main parachute is unsatisfactory. After main parachute deployment, the landing bag is extended to provide attenuation of the landing load. Immediately after landing, the main parachute is automatically disconnected and the reserve parachute is ejected.

The drogue parachute is a 6-foot-diameter conical ribbon-type with a 30-foot-long riser. The main and reserve parachutes are 63-foot diameter ringsail types, either of which will provide a sinking velocity of 30 feet per second at sea level.

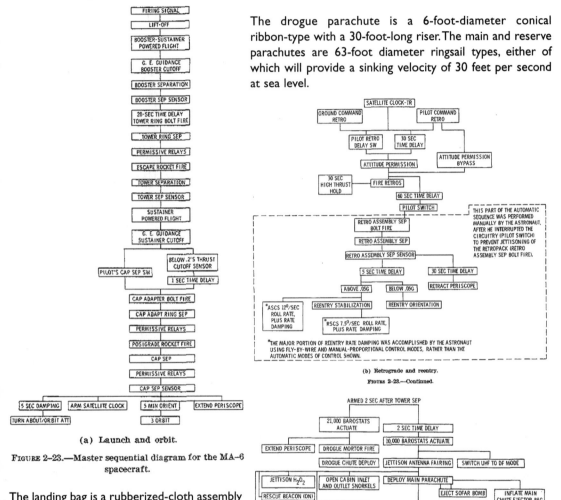

(a) Launch and orbit.

FIGURE 2–23.—Master sequential diagram for the MA–6 spacecraft.

(b) Retrograde and reentry.

FIGURE 2–23.—Continued.

(c) Landing and recovery.

FIGURE 2–23.—Concluded.

The landing bag is a rubberized-cloth assembly about 4 feet long. Before release, the heat shield is held directly to the spacecraft by a mechanical latch and the landing bag is folded and contained between the heat shield and spacecraft. After release, the heat shield drops down and extends the bag to its full length. For a water landing, the bag attenuates landing decelerations from approximately 45g to approximately 15g.

The drogue parachute deployed at an acceptable but somewhat higher than expected altitude during the MA-6 mission. Both drogue and main parachutes were observed by the astronaut to be in good condition after deployment. The heat shield released properly and the landing bag attenuated the landing loads about as expected.

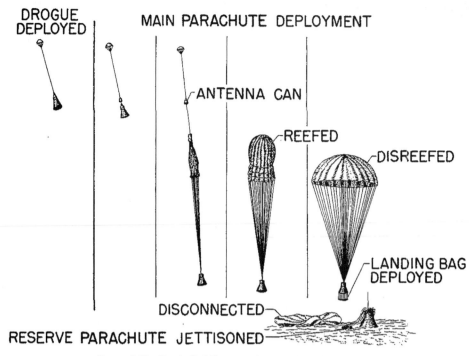

DROGUE
DEPLOYED

MAIN PARACHUTE DEPLOYMENT

ANTENNA CAN

REEFED

DISREEFED

LANDING BAG
DEPLOYED

DISCONNECTED

RESERVE PARACHUTE JETTISONED

FIGURE 2-24.—Sketch depicting use of spacecraft landing systems.

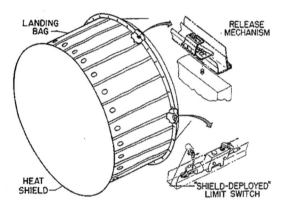

LANDING
BAG

RELEASE
MECHANISM

HEAT
SHIELD

"SHIELD-DEPLOYED"
LIMIT SWITCH

FIGURE 2-25.—Details of heat-shield deployment mechanism and sensing switches.

The "heat-shield-deployed" signal, that is sent to the telemetry system, is furnished by either one of two limit switches which sense the movement of the heat-shield retention devices as shown in figure 2-25. During the orbital flight portion of the MA-6 mission, one of these switches sent a "shield-deployed" signal to the ground monitoring stations. Post flight tests of both limit switches revealed that one switch was faulty and could give intermittent "shield-deployed" signals with the shield locked in place.

FUTURE PLANS

As a result of minor difficulties experienced during the flight of MA-6 the following modifications will be made, to subsequent spacecraft:

(1) The limit switches which indicate that the heat shield is released will be wired in series rather than parallel and rigged farther away from actuation points.

(2) A "maneuver" switch will be installed on the ASCS panel to permit the astronaut to interrupt automatic orbital pitch-precession.

(3) The 1-pound-thrust chamber assemblies are being modified to make them more reliable.

(4) The check valve between the coolant tank and inverter cold plates is to be replaced by a manual valve.

3. LIFE SUPPORT SYSTEMS AND BIOMEDICAL INSTRUMENTATION

By RICHARD S. JOHNSTON, Asst. Chief, Life Systems Division, NASA Manned Spacecraft Center; FRANK H.SAMONSKI, Jr., Life Systems Division, NASA Manned Spacecraft Center; MAXWELL W. LIPPITT, Life Systems Division, NASA Manned Spacecraft Center; and MATTHEW I. RADNOFSKY Life Systems Division, NASA Manned Spacecraft Center

Summary

This report contains a description of the environmental control system and outlines the system performance in flight. The pressure suit is described and the significant pressure suit developments accomplished to date are shown. The Survival kit is described and special emphasis has been placed on newly developed survival items. The MA-6 bio-instrumentation is discussed and the development of the blood pressure measuring system is reviewed. The paper is presented as separate sections for these four areas.

Environmental Control System

Introduction

The Project Mercury environmental control system (ECS) has been described in previous papers (refs. 1 and 2) and, therefore, this paper only reviews the system design and outlines specific MA-6 system configurations. The test program for the ECS was presented in the MR-3 flight report (ref. 3). Flight data for the MA-6 flight are presented in this paper.

System Description

The Mercury environmental control system provides a livable environment for the astronaut in which total pressure, gaseous composition, and temperature are maintained, and a breathing oxygen supply is provided. To meet these requirements a closed-type environmental control system was developed by AiResearch Manufacturing Division of Garrett Corporation under a McDonnell Aircraft Corporation subcontract.

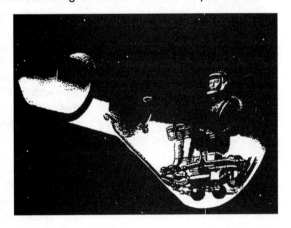

FIGURE 3-1.—Project Mercury environmental control system.

The environmental control system shown in figure 3-1 is located in the lower portion of the spacecraft under the astronaut support couch. The astronaut is clothed in a full pressure suit to provide protection in the event of a cabin decompression.

The pressures in the cabin and pressure suit are maintained at 5.1 psia in normal flight with a 100-percent oxygen atmosphere. The system is designed to control automatically the environmental conditions within the suit and cabin throughout the flight. Manual controls are provided to enable system operation in the event of automatic control malfunction. In describing the environmental control system, it can be considered as two subsystems; the pressure-suit control system and the cabin system. Both of these system operate simultaneously from common coolant water and electrical supplies. The coolant water is stored in a tank with a pressurized bladder system to facilitate weightless flow of water into the heat exchanger. Electrical power is supplied from an onboard battery supply. Oxygen is supplied at an initial pressure of 7,500 psi from two spherical steel tanks.

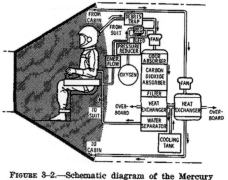

FIGURE 3–2.—Schematic diagram of the Mercury
environmental control system.

Pressure Suit Control System

The pressure-suit control system provides breathing oxygen, maintains suit pressurization, removes metabolic products, and maintains, through positive ventilation, gas temperatures.

As shown in figure 3-2, the pressure suit is attached to the system by two connections, the gas inlet connection at the waist and the gas exhaust at the helmet. Oxygen is forced into the suit distribution ducts, carried to the body extremities, and permitted to flow freely back over the body to facilitate body cooling. The oxygen then passes into the helmet where the metabolic oxygen, carbon dioxide, and water vapors are exchanged. The gas mixture leaves the suit and passes through a debris trap where particulate matter is removed. Next, the gas is scrubbed of odors and carbon dioxide in a chemical canister of activated charcoal and lithium hydroxide. The gas then is cooled by a water evaporative type of heat exchanger which utilizes the vacuum of space to cause the coolant water to boil at approximately 35° F. The heat-exchanger exit gas temperature is regulated through manual control of the coolant water flow valve. The resulting steam is exhausted overboard. The steam exit temperature on the overboard duct is monitored by a thermal switch which actuates a warning light when the duct temperature drops below 47° F.

The light is on the astronaut's panel and provides a visual indication of excessive water flow into the heat exchanger. Proper monitoring of the light and correction of the water flow rate will prevent the heat exchanger from freezing. In the gas side of the heat exchanger, water vapors picked up in the suit are condensed into water droplets and are carried by the gas flow into a mechanical water separation device. The water separator is a sponge device which is squeezed periodically to remove the metabolic water from the system. This water is collected in a small tank. The constant flow rate of the atmosphere is maintained by a compressor.

In the MA-6 spacecraft a constant bleed orifice was provided between the oxygen supply and the pressure suit control system. This constant oxygen flow was in excess of metabolic needs and thus provided a continuous flushing of the pressure suit to insure adequate oxygen partial pressure. In normal operation, suit pressure levels were maintained slightly above cabin pressure by metering this excess oxygen flow through an exhaust port in the demand regulator. In the event of a cabin decompression the demand regulator would automatically establish a referenced pressure of 4.6 psia for the exhaust port of the regulator, and thereby suit pressure would be maintained at this pressure level. The addition of the oxygen bleed orifice is the major ECS change for this flight.

An additional mode of operation is provided by the emergency rate valve. This valve provides an open-type pressure-suit operation similar to aircraft pressure-suit systems. A fixed flow of oxygen is directed through the suit for ventilation and metabolic needs. The remainder is dumped into the cabin. This system is used in the event the pressure-suit control system fails and also during final stages of descent. The other components of the suit system are closed off during this mode of operation.

Oxygen is supplied from two tanks, each containing sufficient oxygen for more than 28 hours. The tanks are equipped with pressure transducers to provide data on the supply pressure. The tanks are connected in such a way that depletion of the primary supply automatically provides for supply from the secondary bottle.

Cabin System

The cabin system controls cabin pressure and temperature. A cabin relief valve controls the upper limit of cabin pressure. This valve permits cabin pressure to decrease with ambient pressure maintaining a differential pressure of 5.5 psi during the climb of the vehicle. This valve seals the cabin at 5.5 psia. In addition, a manual decompression feature is incorporated in this valve to permit the astronaut to dump the cabin pressure if a fire or buildup of toxic gases occurs.

A cabin-pressure regulator meters oxygen into the cabin to maintain the lower limit of pressurization at 5.1 psia. A manual recompression feature is incorporated in the regulator for cabin repressurization after the cabin has been decompressed.

Cabin temperature is maintained by a fan and heat exchanger of the same type as that described in the discussion of the pressure-suit system.

Postlanding ventilation is provided through a snorkel system. At 20,000 feet following reentry, the snorkels open and ambient air is drawn by the suit compressor through the inlet valve. The gas ventilates the suit and is dumped overboard through the outlet valve.

Flight Data

Launch. — The launch phase was normal in that cabin and suit pressures maintained a 5.5 psi differential pressure above ambient during ascent and held at 5.7 and 5.8 psia, respectively.

Orbit. — Cabin and suit pressures were maintained at 5.7 and 5.8 psia, respectively, throughout the flight. The delay in these pressures that has been observed in previous flights was absent in this flight for three possible reasons:

(1) Low cabin leakage (less than 500 cc/min)

(2) Oxygen from the bleed orifice in excess of astronaut requirements

(3) Possible leakage from the secondary oxygen supply

The oxygen partial pressure measurement agreed with suit pressure within 0.2 psi throughout the flight. This value is within the accuracy of the instrument.

The cabin air temperature (fig. 3-3) fluctuated between 90° F and 104° F as the spacecraft passed from darkness into sunlight. The astronaut reported that at least five attempts to reduce cabin air temperature by increasing water flow to the cabin heat exchanger resulted in the illumination of the excess-water light. This light indicated that the cabin heat exchanger was operating near its maximum capacity for the existing conditions. Even so, the mean cabin air temperature was steadily reduced during the mission after the first hour in orbit.

The suit inlet temperature (fig. 3-3) varied between 65° F and 75° F during the orbit phase. The astronaut reported a coolant flow of 1.7 lb./hr to the suit heat exchanger and a steam exhaust temperature of 60° F. These values are both higher than anticipated and contradict each other since freezing of the heat exchanger would be expected at this flow rate. No explanation of this anomaly can be offered at this time.

The coolant tank was charged with 25 pounds of water before the flight. The coolant-quantity indicating system shows a usage of 7.2 pounds. Postflight tests revealed a usage of 11.8 pounds. The difference in calibration and final system temperatures can account for about 3.8 pounds of the 4.6-pound discrepancy. The remainder is

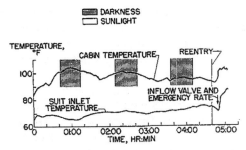

FIGURE 3-3.—Variation of suit and cabin temperatures with time.

considered to be instrument error.

The primary oxygen supply pressure indicates a usage rate of 0.13 lb./hr for the duration of the flight. Postflight tests confirm this usage rate.

The secondary oxygen supply exhibited an unexplained decay in pressure of approximately 12 percent of the total supply. This decay was first noted at an elapsed time of 1 hour and 40 minutes. An approximation of the time when the leakage began is difficult since the bottle was serviced to 8,000 psig prior to flight, and the maximum indicating value of the pressure transducer is 7,500 psig. Postflight testing revealed no appreciable leakage from the secondary supply. No explanation of this problem is available at this time.

Reentry and Postlanding

The maximum cabin temperature during reentry and post landing was 103° F, which was tolerable. The suit inlet temperature increased to 86° F during the post-landing phase. This value is reasonable since the air temperature in the landing area was 76° F (relative humidity, 56 percent) and the suit compressor raises the temperature by approximately 10° F.

Pressure Suit

Introduction

The pressure-suit used in the MA-6 flight was developed from the U.S. Navy MK-IV full pressure-suit manufactured by the B. F. Goodrich Co. This basic suit was selected by NASA in July 1959 for use in Project Mercury after an extensive evaluation program of three full pressure-suits. This initial suit evaluation was conducted by the U.S. Air Force Aerospace Medical Laboratory, Aeronautical Systems Division. Many design changes have been made to the suit since the start of the Mercury program and, indeed, changes and modifications are still being investigated to provide as good a suit as possible for the Project Mercury flights. In this paper the suit is briefly described and emphasis is placed on showing the developmental evolution of the present pressure-suits.

The full pressure-suit consists of five basic components, the suit torso, helmet, gloves, boots, and undergarment.

Pressure-Suit Torso

The suit torso, as shown in figure 3-4, is a closely fitted coverall tailored for each of the astronauts. It covers all of the body except for the head and hands. The torso section is of two-ply construction: an inner gas-retention ply of neoprene and neoprene-coated nylon fabric and an outer ply of heat-reflective, aluminized nylon fabric. The helmet is attached to the torso section by a rigid neck ring. A tie-down strap is provided on this neck ring to prevent the helmet from rising when the suit is pressurized. Straps are also provided on the torso section for minor sizing adjustments of leg and arm length and circumferences and to prevent the suit from ballooning when pressurized.

1. NECK RING
2. HELMET TIEDOWN STRAP
3. PRESSURE SEALING ENTRANCE ZIPPER
4. NECK ZIPPERS
5. WAIST ZIPPER
6. INLET VENT PORT
7. PRESSURE INDICATOR
8. BLOOD PRESSURE FITTING

FIGURE 3-4.—Pressure suit torso.

Donning and doffing of the suit is provided through a pressure-sealing entrance zipper which extends diagonally across the front of the torso from the left shoulder down to the waist. Two frontal neck zippers and a circumferential waist zipper are also provided for ease in donning and doffing.

The pressure-suit ventilation system is an integral part of the torso section. A ventilation inlet port is located at a point just above the waist on the left side of the torso section. This inlet port is connected to a manifold inside the suit where vent tubes lead to the body extremities. These tubes are constructed of a helical spring covered by a neoprene-coated nylon fabric that contains perforations at regular intervals. Body ventilation is provided by forcing oxygen from the environmental control system into the inlet and distributing this gas evenly over the body. The ventilation system in the Mercury pressure-suit was especially developed to insure compatibility with the environmental control system.

The suit torso section contains several items which have been developed specifically for Project Mercury. They are as follows:

Bioconnector.

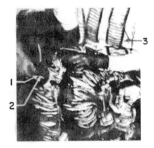

1. INTERNAL PLUG
2. RECEPTACLE PLATE
3. UNDERGARMENT WITH SPACER PATCHES

FIGURE 3–5.—Bioconnector (installation).

The bioconnector provides a method for bringing medical data leads through the pressure suit. The bioconnector consists of a multipin electrical plug to which the biosensors are permanently attached, a receptacle plate mounted to the suit torso section and an outside plug which is connected to the spacecraft instrumentation system. With this system, the biosensor harness is fabricated with the bioconnector as an assembly and no additional electrical connectors are introduced into the transducer system. In operation (fig. 3-5) the male internal plug is inserted inside the suit receptacle and locked into place. The internal plug protrudes through the suit to allow the spacecraft plug to be attached. The bioconnector system has proven to be a much more satisfactory connector than the previously used biopatch.

Neck Dam.

A conical rubber neck dam is attached to the torso neck ring as shown in figure 3-6. The purpose of this neck dam is to prevent water from entering the suit in event of water egress with the helmet off. The neck dam is rolled and stowed on the outside of the neck ring disconnect. After the astronaut removes the helmet in preparation for egress, he unrolls the neck dam until it provides a seal around his neck.

Pressure Indicator.

A wrist-mounted pressure indicator is worn on the left arm. This indicator provides the astronaut a cross check on his suit pressure level. The indicator is calibrated from 3 to 6 psia.

Blood-Pressure Connector.

A special fitting is provided on the suit torso which permits pressurization gas to be fed into the blood-pressure cuff. A hose leading from the cuff is attached to this connector during suit donning. After astronaut ingress into the spacecraft the pressurization source is attached to the connector on the outside of the suit.

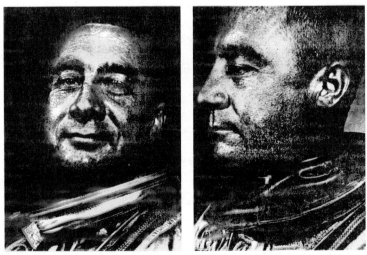

(a) Stored position.

(b) Unrolled position.

FIGURE 3–6.—Neck dam.

Helmet

The helmet assembly, shown in figure 3-7, consists of a resinous, impregnated Fiberglas hard shell; an individually molded crushable impact liner; a ventilation exhaust outlet; a visor sealing system; and a communications system.

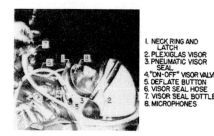

1. NECK RING AND LATCH
2. PLEXIGLAS VISOR
3. PNEUMATIC VISOR SEAL
4. "ON-OFF" VISOR VALVE
5. DEFLATE BUTTON
6. VISOR SEAL HOSE
7. VISOR SEAL BOTTLE
8. MICROPHONES

FIGURE 3-7.—Helmet assembly.

The helmet visor scaling system consists of a pivoted Plexiglas visor, a pneumatic visor seal, and an on-off visor valve. Closing the visor actuates the valve and causes automatic inflation of the visor seal. The visor seal remains inflated until a deflation button on the valve is manually actuated by the astronaut. The valve has provision for attachment of the visor-seal gas supply bottle hose.

The helmet communication system consists of two independently wired AIC-10 earphones with sound attenuation cups and two independently wired AIC-10, newly developed, dynamic, noise-cancelling microphones. The microphones are installed on tracks which allow them to be moved back from the center of the helmet to permit eating and proper placement.

Gloves

The gloves attach to the suit torso at the lower forearm by means of a détente ball-bearing lock. The gloves have been specially developed for Project Mercury to provide the maximum in comfort and mobility. Early centrifuge programs dictated the requirements for this development. Poor mobility in wrist action when the suit is pressurized caused an impairment in the use of the three-axis hand controller.

A pressure sealing wrist bearing was incorporated to improve mobility in the yaw control axis. The one-way stretch material on the back of the gloves improves mobility in the pitch and roll axes.

The gloves have curved fingers so that when pressurized the gloves assume the contour of the hand controller. The glove, like the torso section, has a two-ply construction — the inner gas retention ply and an outer restraint ply. The inner ply is fabricated by dipping a mold of the astronaut's hand into Estane material. The outer ply is fabricated from one-way stretch nylon on the back of the hands and fingers and a neoprene material injected into a nylon fabric in the palm of the gloves to prevent slippage in turning knobs, and so forth. Lacings are provided on the back of the glove to allow for minor adjustments. Two wrist restraint straps are provided to form break points and thereby improve pressurized glove mobility.

Miniature needle-like red finger lights are provided on the index and middle fingers of both gloves. Electrical power is supplied to the miniature lights by a battery pack and switch on the back of the gloves. These lights provide instrument-panel and chart illumination before the astronaut is adapted to night vision.

Boots

Lightweight, aluminized, nylon-fabric boots with tennis shoe type soles were specially designed for the Mercury pressure suit. These boots resulted in substantial weight savings, provided a comfortable boot for flight, and a flexible friction sole which aids in egress from the spacecraft.

Undergarment

The undergarment is a one-piece, lightweight, cotton garment with long sleeves and legs. Thumb loops are provided at the sleeve ends to prevent material from riding up the arms during suit donning. Ventilation spacer patches (see fig. 3-5) of a trilock construction are provided on the outside of the undergarment to insure ventilation gas flow over certain critical areas of the body.

Pressure - Suit Support

Prior to and after astronaut donning of the pressure-suit the complete assembly was pressurized and leak checked at 5 psig, and at 5 inches of water differential pressure. This test console provides the pressure control and leakage measurement system required.

During astronaut transfer from the suit dressing room to the launching pad, a lightweight, hand-carried, portable ventilator provided suit cooling. Constant communications are maintained with the astronaut during this transfer by utilizing portable communication headsets carried by the astronaut insertion team.

In the MA-6 flight, the pressure suit served more as a flight suit since the cabin pressure was maintained. Astronaut comments indicated that the pressure suit was satisfactory throughout the flight.

Survival Equipment

Introduction

The MA-6 spacecraft was equipped with a survival kit (fig. 3-8) made up of standard Department of Defense (DOD) survival items and other items recently developed by NASA. This survival equipment is carried for emergency recovery contingencies and has not been used in the three manned flights to date.

Contents

Contents of the survival kit are shown in figure 3-9 and are as follows -.

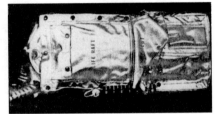

FIGURE 3-8.—Packed survival kit.

Sea dye marker
Survival flashlight
Shark chaser
Food container
Jack knife
Sun glasses
Pocket waterproof matches
Signal whistle
Survival knife

Signal mirror
Zinc oxide
Soap
Medical injectors
First-aid kit
SARAH beacon
Nylon lanyard
Life raft
Water container

The newly developed items include the flashlight, life raft, water container, and automatic medical self injectors. Also, a newly developed life vest, not contained in the survival kit, was developed. This report presents a brief description of each of these, items and the developmental programs.

Life raft

Early in the Mercury program it was decided that improvements could be made in the design of the PK-2 raft. These improvements included improved stability in rough seas and ease in boarding. A contract was let in November 1959 to fabricate several life rafts which incorporated bow ballast buckets under the raft for stability and a deflatable aft section to simplify boarding. Tests of these rafts with a subject in a full pressure suit proved that they were very difficult to capsize and easy to board.

With the configuration settled upon, an attempt was made to reduce the weight of the raft by the employment of a lighter base fabric and coating and CO_2 cylinders.

Sea tests were again performed by NASA personnel attired in a Mercury pressure-suit using life rafts fabricated of conventional fabric, built in accordance with the NASA designs. The subject was able to sit on one side tube of the raft without capsizing it.

An in house program was then initiated to improve the raft reliability and reduce its weight by decreasing the number of fabrication seams through the utilization of new materials. Figures 3-10 and 3-11 are photographs of the rafts developed by NASA.

Test results indicated satisfactory or superior performance when compared with conventional rafts or rafts developed to date for NASA. The new unit was significantly lighter than conventional rafts, packed to about 1/4 the thickness, and when inflated provided over 3 square feet of additional space for the occupant. This raft contained only 1 seam as opposed to 11 seams in the previous rafts.

Two single-seam rafts were fabricated and strength tested to 5 psi and shape retention after 24 hours with 2 psi. These rafts were then packed and subjected to the shock, acceleration, temperature, vibration, vacuum, and oxygen conditions specified for the Mercury spacecraft. After passing these tests, the rafts were re-inflated, repacked, and considered flight items for the MA-6 mission.

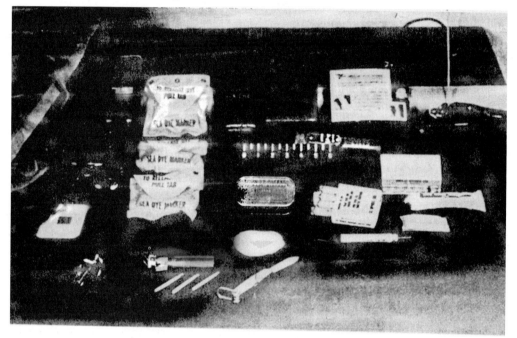

FIGURE 3-9.—Survival kit components.

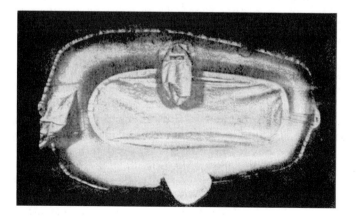

FIGURE 3-10.—Liferaft showing stabilizing buckets.

Water Container

The original spacecraft water containers were two inflexible one pound plastic cases containing 3 pounds of water each, with a total weight of 8 pounds. It was determined that a flexible water bag might be installed in the life raft kit which would provide both in-flight and survival drinking water and reduce overall weight and volume.

By fabricating these items of neoprene coated nylon fabric, an immediate savings of 1¾ pounds and 220 cubic inches in volume was realized, since when filled with 6 pounds of water, the bag takes up space already unusable in the life raft kit.

Figure 3-12 shows the water container in the un-inflated, unfilled condition. Water is forced under pressure into the container by means of the one-way pressure valve shown in the lower, middle, left-hand section. The astronaut drinks through the plastic, spiral tube. An Estane liner within the bag insures tasteless water. Figure 3-13 shows how the water container packs into the survival kit. The life raft is placed on top of the water container.

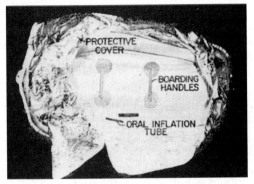

FIGURE 3-11.—Liferaft top view.

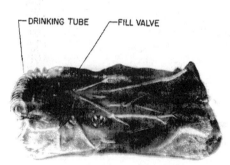

FIGURE 3-12.—Water container.

FIGURE 3-13.—Water container in survival kit.

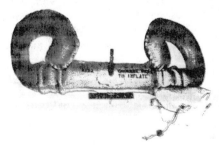

FIGURE 3-14.—Life vest.

Life Vest

As a result of the MR-4 recovery operation in which the astronaut had to make an emergency exit without his survival kit, it became apparent that an emergency flotation device was required to maintain flotation in a water-filled pressure-suit. In order to preclude or reduce the possibility of similar situations of this nature, NASA instituted the development of a miniaturized life vest employing the following criteria:

(1) Minimal bulk (less than 20 cu in.)

(2) Minimal weight (less than 1 lb.)

(3) Minimal interference with flight efficiency

Two basic configurations were fabricated, one a simple tube similar to a ski belt, the second a belt with inflated hooks (fig. 3-14). The inflatable hooks were constructed to provide a positive grasp to the wearer's shoulders; easy donning and doffing features; and adequate flotation characteristics without impairing rescue, recovery, or swimming of the astronaut.

The second model, with minor modification and supplementary testing, became the standard flight items of equipment.

Internal carbon dioxide actuating devices were designed to reduce bulk and weight. Lighter weight coated fabrics were tested and one was selected (5 oz nylon with 1.5 oz neoprene).

A packet to stow the vest was developed. The present configuration (fig. 3-14) is trapezoidal, 5 by 4 by 3 by 1 inch in thickness. The packet and vest weigh less than 1 pound (0.99 lb.) and contain an oral inflater in addition to the internal carbon dioxide charge. Presently, the packet is affixed to the suit below the neck ring. A lanyard is provided to preclude accidental loss upon inflation. Placement tests of the package indicate the chest area to be as satisfactory as the lower leg. The entire unit can be opened, inflated, and donned with one hand in less than 10 seconds when attired in the pressure suit.

Tests were made in the open sea from a launch. The test subject first swam about with the pressure suit in the intact condition, then actuated the flotation device, donned it, and opened the zipper to his suit. The suit soon filled with water and the subject swam about unhindered. The subject then was instructed to remove the vest while tied to a safety line. He was unable to remain on the surface unless held there by the safety line.

Final acceptance testings, including high acceleration, heat-cold, vacuum, and were performed in accordance with the requirements for all spacecraft hardware. The life vest was carried by the astronaut during the MA-6 flight but was not used.

Automatic Self-Injectors

The survival kit contained four automatic self-injectors which contained medications for pain, shock, and motion sickness and a stimulant. These injectors were developed under a NASA contract The injectors are stored in a small package. One end of the injector is equipped with a red safety cap and the other end contains the medication and needle. Upon removal of the safety pin, the injector is armed. By pressing the needle end of the injector into the pressure suit, the needle is extended through the suit into the skin and the medication is released. The resulting hole in the suit caused an insignificant suit leak. In the MA-6 flight the astronaut did not use any of the injectors prior to, during, or after the flight.

Post Recovery Kit

Each of the MA-6 recovery ships was equipped with a post recovery kit which contained:

High cut gym shoes	Toothbrush	Flight jackets	Razor and blades and cream
T-shirt	Socks	Shorts (briefs)	CombsWrist watch
Sunglasses	Handkerchiefs	Soap	Pressure-suit-helmet carrying case
Wash cloths	Postflight coverall		

Bio-instrumentation

Introduction

The biosensors used to monitor the physiological state of the pilot during the MA-6 flight are essentially the same as those used on the previous MR-3 and MR-4 suborbital flights with the exception that for the first time the astronaut blood-pressure measuring system (BPMS) was used. (See fig. 3-15.)

A detailed discussion of the electrocardiogram, body temperature, and respiration sensors can be found in the report on the MR-3 flight (ref. 3), and these sensors will be only briefly treated here. A more complete discussion of the BPMS is included.

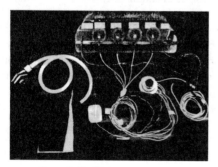

FIGURE 3-15.—Flight sensor harness.

Electrocardiographic Sensor

The ECG sensors consist of rings composed of silicone rubber. The rings are constructed to support a disc of 40-mesh stainless steel screen, 30 mm in diameter and approximately 2 mm above the skin. The center conductor of a miniature-type coaxial cable is brought through a strain-relieving projection in the rubber ring and soldered to the screen. A piece of thermally shrinking plastic tubing seals the cable shield at the entrance into the ring to prevent the entry of moisture.

Before the electrode is applied to the washed and shaved

skin, a coating of elastoplast adhesive is applied to the bottom surface of the electrode and allowed to dry. The ring cavity is filled with a paste composed of bentonite, calcium chloride, and water. The electrode is then applied and the cavity of the ring is checked; voids are eliminated; then the assembly is sealed with tape. A 4-inch square of moleskin applied over the entire sensor area completes the installation.

The signal from the ECG electrode is transmitted via the coaxial cable within the pressure suit to the bioconnector and then to the ECG amplifiers in the spacecraft instrumentation package. Differential amplifiers with high input impedance and good common mode rejection are used to raise the signal to that required for input to the spacecraft telemetry system. ECG measurements are discussed in paper 9.

Respiration Sensor

In order to measure respiration rate and depth, a thermistor anemometer detecting the flow of the expired air was used. A thermistor with sufficient current through it to maintain its temperature at approximately 200° F in still air was mounted in a small plastic enclosure that was attached to one of the microphones within the helmet. A funnel-shaped opening facing the pilot conducts a portion of the expired air across the heated thermistor, and through exit vents in the back. When the flow of air cools the thermistor, the resistance change causes a voltage variation across the sensor. This voltage change is sensed by a small preamplifier mounted on the cable leading from the sensor and this signal is transmitted through the pressure-suit biconnector to the spacecraft instrumentation package.

The respiration sensing system does not yield data from which tidal volume can be determined. The microphone to which the sensor is attached is pivoted so that it can be adjusted. When the microphone is moved, the signal from the sensor varies because of the change in the volume of air passing across the thermistor. The respiration data have not been fully satisfactory to date. Therefore, an improved respiration sensing system employing the impedance pneumograph principle is presently being developed under a NASA Manned Spacecraft Center contract. Details of the impedance pneumograph are published in reference 4.

Body Temperature

The body temperature probe is a thermistor mounted in a special rectal catheter. The catheter is a small plastic cylinder about 3 mm in diameter and 25 mm long from which the thermistor projects approximately 2 mm. The thermistor, catheter, and lead wires are dipped in liquid latex to a length of about 20 cm to prevent the entry of moisture. The thermistor forms one arm of a resistance bridge which is excited by 400 cps current and which is located in the spacecraft instrumentation package. Body temperature data are discussed in paper 9.

Blood - Pressure Measuring System

In April 1961 it was decided to institute a program to develop a device to measure arterial pressure. It was hoped that a system could be developed in time for the first orbital flight. After a survey of the current state of the art, the decision was made to use the new method then currently under development. An intensive effort to design, develop, and test the flight hardware was started in June 1961. The method utilizes essentially the same principle used in clinical sphygmomanometry, namely an inflatable occluding cuff on the left arm. The cuff is inflated by gas to a pressure in excess of expected systolic pressure. As the pressure decreases slowly, a microphone placed under the lower half of the cuff over the brachial artery transduces the Korotkoff sounds (ref. 5). The signal from the microphone is amplified and mixed with a signal from a pressure transducer which transmits the cuff pressure. In order to find the arterial pressure it is necessary to identify the points of inception and cessation of the microphone signal on the cuff pressure signal, which are the systolic and diastolic pressures.

In order to develop a blood pressure measuring system (BPMS) for spacecraft use, a number of problem areas had to be considered:

(1) Pilot safety and comfort

(2) Establishment of the accuracy of the measurement compared to clinical and direct methods

(3) Operation in a full-pressure suit

(4) Operation on an active subject in a noisy environment

(5) Compatibility with spacecraft systems

(6) Compatibility with the receiving facilities at Mercury Control Center and the Mercury Network stations

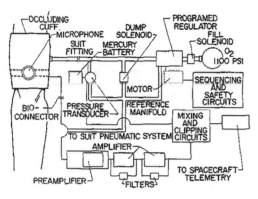

FIGURE 3–16.—Block diagram of automatic BPMS.

The original concept of the BPMS (fig. 3-16) was an automatic system, which would be initiated from a tracking station through the command receiver, by an automatic sequencing device onboard or by the pilot. The automatic system incorporated special safety circuits to dump the cuff pressure if the cuff stayed above 60 mm Hg for more than 2 minutes. This feature provided for the situation in which the pilot was unconscious and the automatic system failed to bleed off the cuff pressure. The cuff pressure in the automatic system was decreased in a linear manner from 220 mm Hg to 60 mm Hg by a special pressure regulator in which the reference spring tension was varied by a motor-driven cam.

The pneumatic system consisted of an oxygen storage flash, solenoid fill valve, motor-driven regulator, dump solenoid valve, cuff pressure transducer, and suit reference manifold. The regulator, dump solenoid valve, and pressure transducer were referenced to a manifold connected through a flow restrictor to the pressure-suit system to prevent differences between cabin pressure and suit pressure from causing large errors and to allow measurements in the event of the loss of cabin pressure. After considerable testing it was decided to wear the cuff inside the suit because readings taken with the cuff outside of the suit showed large errors due to the cooling ducts within the suit. The problem of entering the suit with a pneumatic line to inflate the cuff required extensive development. A fitting was devised that is comfortable, reliable, and easily disconnected for spacecraft egress.

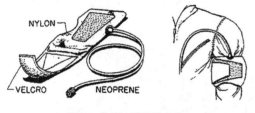

FIGURE 3–17.—BPMS cuff.

During the testing of the preliminary system on astronauts and other flight personnel it became evident that the standard 5-inch clinical cuff was unsatisfactory because of its stiffness, bulk, and the fact that arm movement was restricted. A new cuff (fig. 3-17) was devised by NASA which has proved to be acceptable on all points and is almost unnoticeable in its un-inflated state. Tests were performed to compare the new cuff with the standard cuff and the resulting data were identical. It is felt that the new cuff type has application where comfort, light weight, ease of application and unrestricted arm motion are desired.

Dr. Geddes observed in reference 5 that if the microphone signal is filtered so that only frequencies between 32 and 40 cps are used, the various artifacts due to movements and ambient noise level are greatly attenuated, while the component that allows the discrimination of the systolic and diastolic points is passed. Sometime prior to the inception of the project, this observation was confirmed. For flight use a specially damped, piezoelectric microphone was developed. The instrument is about 3.5 cm in diameter and 0.5 cm thick and is so constructed that sensitivity to noise entering from the side away from the skin is greatly reduced. The microphone signal exits from the suit through the bioconnector and enters the amplifier in the blood-pressure unit. The BPMS amplifier consists of a shielded preamplifier and two high-gain amplifiers which determine the response characteristics. Each amplifier is designed to have greatly attenuated response outside the 32 to 40 cps pass band by means of resistor-capacitor filtering circuits in each feedback loop. The amplifier output is gated so that unless a signal of sufficient amplitude is present there is no output signal, and this gating results in a marked reduction in the output noise level for improved readability of the signal.

The cuff pressure is measured by a potentiometer-type transducer powered by two mercury batteries to give the zero-centered +1.5 volt output necessary for input to the telemeter. The signal from the pressure transducer passes through a miniature transformer where it is mixed with the output from the microphone and then on to the output clipping circuits that protect the telemetry system from excessive voltages that can cause cross-channel interference.

In order to compare this method with direct arterial measurement, a special centrifuge unit was fabricated and installed on the human centrifuge at the University of Southern California (USC). A series of tests were performed by personnel from USC, NASA, McDonnell, and AiResearch. Subjects equipped with the BPMS on the right arm and an arterial catheter on the left arm were tested at various acceleration levels. Spot checks were also made with a clinical cuff and stethoscope. The results showed that at 1g the BPMS method read about 5 mm lower on systole and about 5 mm higher on diastole compared with the direct arterial readings.

There is an increased scatter in the points as acceleration increased which is thought to be partly due to the "eyeballs-down" position of the subject causing a pooling of blood in the lower arm. Comparison of the data. from the BPMS with clinical and arterial tests can be summarized as follows:

The BPMS is more accurate than the clinical method when both are compared with the direct arterial measurements, and the BPMS readings compared with the clinical readings are higher on systole and lower on diastole, a fact which is probably due to the increase in sensitivity of the microphone over the stethoscope.

In order to test the system further as well as to obtain baseline data on the pilots, a centrifuge unit was installed on the centrifuge at the U.S. Navy Aviation Medical Acceleration Laboratory (ANAL), Johnsville, Pennsylvania. Various noise and vibration problems were encountered and solved. The tests were most useful in the testing of the proposed flight amplifier, and they also provided the first opportunity to obtain pilot comments on the device. It was during these tests that the pressure-suit fittings and special cuffs were developed.

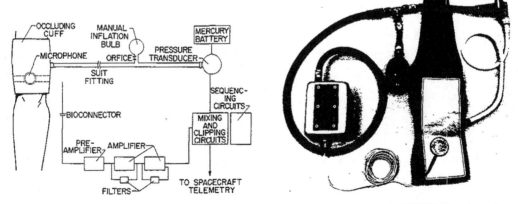

FIGURE 3–18.—Block diagram of manual BPMS.   FIGURE 3–19.—Manual BPMS flight equipment.

Concurrently with the test program, problems in spacecraft integration were being pursued by the McDonnell Aircraft Corporation. The decision to start the program to develop the BPMS as late as 1961 resulted in the system being a retrofit item instead of planned spacecraft equipment. A number of late changes were made in the configuration of the system to reflect developments in the spacecraft equipment. The area originally selected for the mounting of the BPMS was proved undesirable due to egress difficulty.

Alternate areas selected required repackaging of various components and finally resulted in the elimination of the gas pressure source, regulator, and motor programmer and the installation of a hand pumped inflation system with a simple orifice to release cuff pressure (figs. 3-18 and 3-19).

Subsequent flights will be provided with a BPMS requiring only a switch actuation to initiate the cycle. This system will contain the gas pressure source for cuff inflation, the regulator, and the orifice to relieve cuff pressure and will allow a larger number of determinations.

In order to measure arterial pressure without adding telemetry channels, the input to the channel carrying the sternal ECG lead was switched to BPMS during blood-pressure determinations. The band width required for the BPMS is somewhat greater than that required for ECG and it was necessary to modify the receiving equipment to insure legible readout. Blood pressure measurements are discussed in detail in paper 9.

References

1. JOHNSTON, RICHARD S.: Mercury Life Support Systems, Life Support Systems for Space Vehicles. (Presented at the IAS 28th Annual Meeting, New York, Jan. 25-27, 1960), SMF Fund Paper No. FF-25.

2. GREIDER, HERBERT R., and BARTON, JOHN R.: Criteria for Design of the Mercury Environmental Control System — Method of Operation and Results of Manned System Operation. Jour. Aerospace Med., Vol. 32 no. 9, Sept. 1961, pp. 839-843.

3. WHITE, STANLEY C., JOHNSTON, RICHARD S., et al,: Review of Biomedical Systems for MR-3 Flight: Proc. Conf. on Results of the First U.S. Manned Suborbital Space Flight, NASA, Nat. Inst. Health, and Nat. Acad. Sci., June 6, 1961, pp. 19-27.

4. GEDDES, L. A., HOFF, H. E., HICHMAN, D. M., and MOORE, A. G.: The Impedance Pneumograph. Jour. Aerospace Med. vol. 33, no. 1, Jan. 1962, pp. 28-33

5. GEDDES, L. A., SPENCER, W. A., and HOFF, H. E.: Graphic Recording of the Korotkoff Sounds. American Heart Jour., Vol. 57, no. 3, Mar. 1959, pp. 361-370.

 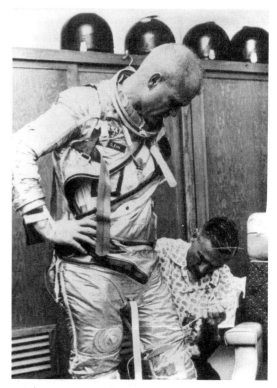

4. LAUNCH-COMPLEX CHECKOUT AND LAUNCH-VEHICLE SYSTEMS

By B. PORTER BROWN, Mercury Launch Coordinator, NASA Manned Spacecraft Center; and G. MERRITT PRESTON, Chief, Preflight Operations Division, NASA Manned Spacecraft Center

Summary

In summary this paper has pointed out the planning required to support the launch complex and vehicle for the MA-6 operation. Fortunately, all modification and emergency type considerations were not activated during the operation. However, the paper has indicated the necessity for such items in support of manned spacecraft operation. It is not the intent of this paper to suggest that all possible combinations of occurrences were thought of and planned for. Such a conclusion can only be reached after considerable experience is gained from many such operations. The success of MA-6, however, indicates that the planned concept, test procedures, and check-out and preparation techniques were sound and that no additional major modifications are necessary for support of a manned orbital operation.

Introduction

This paper is concerned with the special modifications and considerations for Mercury launch operation involving the launch complex and the launch vehicle. The paper covers two areas, long-range planning, and the tests and preparations of the complex and the launch vehicle for the MA-6 operation. For the sake of clarity, comparisons are made between standard Atlas boosters and complexes and the launch vehicle and complex as configured for Mercury.

Description of Launch Complex and Launch Vehicle

The Mercury-Atlas 6 vehicle-was launched from launch complex 14 at Cape Canaveral, Fla. The launch vehicle used for this mission was essentially an Atlas (series) D and the launch complex was basically designed to support Atlas D operations. Both the complex and the launch vehicle, however, were modified to provide various and specific features that were necessary for a manned spacecraft operation. A brief description of a standard launch complex and launch vehicle is given first so that the special features for Mercury will be more readily recognized. Figure 4-1 shows a standard Atlas D launch complex. The term "complex" includes such facilities as the blockhouse, fuel and liquid oxygen storage, electrical power supply, service tower, and the launching pad. All equipment necessary to check out completely each system on the complex and in the launch vehicle is located in these facilities and each of these systems is completely validated prior to each launch operation. Figure 4-2 shows a view of some of the checkout equipment located inside the blockhouse.

The general configuration of the launch vehicle is shown in figure 4-3. The launch vehicle is a 1½ stage, liquid propellant launch vehicle with five engines: 2 booster engines, 1 sustainer engine, and 2 small vernier engines. These engines produce a total thrust of approximately 360,000 pounds. The fuel tank is located immediately above the main engines and the liquid oxygen tank is located above the fuel tank, the tanks being separated by a bulkhead. System components, such as command receivers, telemetry packages, guidance equipment, antennas, and so

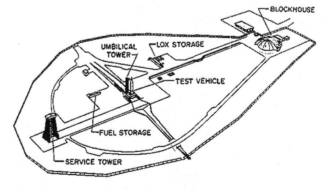

FIGURE 4-1.—Standard Atlas D launch complex.

forth, are housed in the two pods on the sides of the fuel tank. Launch-vehicle guidance is provided by a combination of onboard equipment and radio ground guidance equipment.

Perhaps the best way to explain the normal functions of the launch vehicle is to look at the sequence of events during powered flight. Such a sequence is shown in figure 4-4. The figure shows the launch-vehicle trajectory plotted as altitude against range. Two seconds after liftoff, the roll program is initiated by onboard flight equipment. This maneuver is necessary because the launch pad is oriented so that the pitch axis of the launch vehicle is aligned on an azimuth of 105° while the Mercury spacecraft insertion head is about 75°. Therefore, the roll program has to rotate the launch vehicle approximately 30° to align it with the Mercury spacecraft insertion heading. At 15 seconds after lift-off the roll

FIGURE 4-2.—View of equipment in blockhouse.

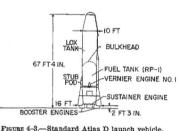

FIGURE 4-3.—Standard Atlas D launch vehicle.

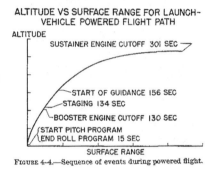

FIGURE 4-4.—Sequence of events during powered flight.

program is complete and then the pitch program is started. Although the pitch program is active throughout the remaining portion of powered flight, the rate at which the pitch program changes the pitch attitude varies throughout the trajectory. At T+130 seconds, the ground guidance station sends a command which shuts off the two booster engines. Then the sustainer engine is locked in the neutral position and the booster engines are jettisoned. The sustainer engine is then unlocked and the vehicle is guided to insertion by the controllable sustainer engine which is positioned by commands from the ground guidance station. At approximately T+300 seconds, the ground guidance station, once satisfied that all insertion parameters are attained, sends a command which shuts off the sustainer and vernier engines.

<u>Complex and Launch-Vehicle Modifications</u>

As mentioned previously, in order to support Mercury missions, both the complex and the launch vehicle were modified. Figure 4-5 shows the major modifications that were made to the complex. In the service tower, a room was built to enclose the spacecraft. Figure 4-6 shows an external view of the service tower and the specially built room. This room, commonly called the "white room," is located near the top of the service tower. The spacecraft is shown in the figure immediately outside of the sliding doors of the white room. Figure 4-7 presents a close-up view looking into the white room. The figure shows the sliding doors in the open position and the spacecraft suspended just above the adapter. The figure also shows the roof in the folded position, but the floor is shown intact. It should be pointed out that the floor also can be folded in a manner similar to that of the roof. The movable doors, floors, and roof are necessary to allow opening of the white room so that the service tower can be moved away from the flight vehicle approximately 55 minutes prior to launch. The environment in this white room was controlled to minimize the effects of humidity, dust, and so forth, on the spacecraft components.

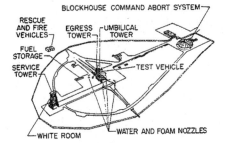

FIGURE 4-5.—Mercury modifications to launch complex.

FIGURE 4-6.—Service tower showing white room.

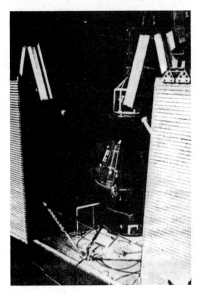

FIGURE 4-7.—Inside view of white room.

An emergency egress tower is shown in figure 4-8. The figure shows the egress platform in the extended position such that the end of the platform is adjacent to the door of the spacecraft. When retracted, the platform is rotated in the vertical plane about the opposite end and locked in the vertical position. This feature is necessary so that the launch vehicle, when launched, will not strike the platform. Actually, the platform is held in the vertical position during the entire countdown and, if needed, it is lowered to the extended position in about 30 seconds by means of remote control from the blockhouse. This tower provided the astronaut with a means of evacuating the spacecraft without external aid or, in case the astronaut became incapacitated, the external egress crew could use the tower to remove the astronaut. Also shown in the figure is the mobile egress tower, known as the "cherrypicker." The mobile tower is shown on the left of the figure in a partially extended position. This mobile tower was used on the Mercury-Redstone operations; however, subsequent tests on the Atlas complex indicated that the tower may possibly interfere with radio transmissions. Also, the tower was subject to possible damage from the greater pressure environment produced by the Atlas engines. It was decided, therefore, that the mobile tower was not as well suited for an Atlas launch as was the fixed structure previously discussed. The "cherrypicker" however was stationed behind the blockhouse so that it could be used as substitute in the event that the primary egress tower failed to operate.

Special rescue and firefighting vehicles were stationed just outside of the complex to transport the egress crew to the tower and/or to meet the astronaut at the tower and transport him away from the complex. Figure shows the position of these vehicles relative to the launch pad. The astronaut-transport vehicle, with its covering of special thermal insulation, can be seen in this figure. The egress procedure was practiced many times and it is interesting to note that the astronaut could evacuate the spacecraft and be delivered to a safety zone outside of the complex in about 2½ minutes.

FIGURE 4-8.—Emergency egress tower.

FIGURE 4-9.—Rescue and firefighting vehicles.

A special firefighting system was also installed. The four nozzles shown in figure 4-5 were remotely controlled from the blockhouse in such a manner that water or fire-smothering foam could be directed to any area inside the complex. Figure 4-10 shows a close-up view of one of the nozzles.

A radio command system was installed in the blockhouse. This system provided a ground command means of firing the spacecraft escape rockets and aborting the spacecraft prior to launch and was the primary system for abort during the first 10 seconds of flight.

FIGURE 4-10.—Firefighting nozzle.

FIGURE 4-11.—Mercury configuration of flight vehicle.

The Mercury configuration of the flight vehicle is shown in figure 4-11. The adapter which mates the spacecraft to the launch vehicle is immediately above the liquid oxygen tank and the spacecraft is mounted on top of the adapter.

Pilot Safety Program

It was recognized at the beginning of the program that the launch vehicle would have to be modified in some areas for Project Mercury therefore a special study program was initiated to evaluate each system, concept of operation, and the effects of combinations of various failures that could conceivably occur. This program, called the "Pilot Safety Program," drew on the talents of many groups, Primarily those groups with previous experience gained from Atlas D operations. The philosophy stressed in this program was based on the use of fully developed components in order to preserve system reliability as established by flight experience. The program also established a standard for components to be used on Mercury launch vehicles so that component acceptability could be based on nominal performance characteristics rather than outstanding or better than expected characteristics. There were some instances in which wiring or circuitry changes were made, but in these cases the changes were made to improve system reliability. The factory rollout procedures and the flight safety review for the pilot safety program are discussed in detail in references 1 and 2.

No attempt is made to mention all changes made to the launch vehicle; however, the major changes will be discussed. For instance, after ignition, the launch vehicle is intentionally held down for several seconds in order to determine that the engines are functioning properly. This change was a result of previous experience which showed that, after ignition, the engine performance could possibly become erratic (rough combustion) and cause destruction of the launch vehicle. The experience also showed that the additional holddown time would provide sufficient time to detect such a malfunction and shut off the engines before lift-off, thereby preventing destruction.

Another system that was modified is the command destruct system. This system was changed to include a time delay circuit so that if a manual destruct command was sent to the launch vehicle, receipt of the command would immediately fire the spacecraft escape rocket motor; but destruct action of the launch vehicle would be delayed 3 seconds to allow the spacecraft time to escape from the launch vehicle.

The addition of the time delay circuit introduced the major modification made to the launch vehicle — the abort sensing implementation system (ASIS). This system was designed specifically for Project Mercury, and its purpose was to provide an automatic system that would sense specific quantities in the launch vehicle, detect when those quantities indicated impending catastrophe in the launch vehicle, and abort the spacecraft to escape the catastrophe. It was believed that the ASIS was necessary because some previous flights of the Atlas D had indicated that the time period between an indication of impending catastrophe and launch vehicle destruction could be extremely short — approaching the reaction time of a human being. It was decided therefore that an automatic system would be desirable, at least until more experience was gained on manned flights.

Basically, the ASIS consists of sensing elements which detect malfunctions and a control unit that receives the signal and initiates the proper action. The sensors used in this system are rate gyros, pressure switches, and electrical power sensors. The control unit is basically connected to two systems, the spacecraft escape system, and the booster-engine system. For example, if the control unit receives a signal from a sensor, the unit tells the spacecraft escape system to abort the spacecraft; then the unit tells the booster engines to shut down. Also, if the engines are intentionally shut down by a command from ground control, the control unit calls for spacecraft abort. It should be pointed out that just prior to launch, the ASIS indicates to the blockhouse that the system is in a ready condition; however, the system is not actually activated until the launch vehicle has risen 2 inches. This feature precludes spacecraft abort in the event that the engines should shut down after ignition but prior to launch-vehicle release. Of course, the ASIS is far more complicated than implied by this discussion, however the intent of this paper is to present the general explanation of operation rather than details of the system.

FIGURE 4-12.—Launch vehicle being erected.

Systems Preparation for MA-6

All of the previous discussion has dealt with long-range planning and implementation in regards to the complex and the launch vehicle for the Mercury program. The following discussion will concern the actual preparation of these systems for the launch of MA-6. Upon arrival at Cape Canaveral, the launch vehicle was inspected and prepared for erection in approximately 48 hours. The launch vehicle was transported to the complex on a dolly-type vehicle. The launcher on the launch pad was rotated about 90°; and the launch vehicle was backed into the launcher, aligned and attached to the launcher. A hoist cable was then attached to the front end of the dolly (top end of the launch vehicle) and the dolly and launch vehicle were hoisted to the vertical position, the launcher rotating back to its original position. Figure 4-12 shows the launch vehicle being erected, and figure 4-13 shows the launch vehicle after erection in launch position.

FIGURE 4-13.—Launch vehicle after erection.

After erection it was learned that the launcher mechanism, in which the launch vehicle was mounted, could not be adjusted sufficiently to align the launch vehicle properly. Therefore, the launch vehicle was taken down, the launcher mechanism was replaced, and the launch vehicle was re-erected. All systems on the complex and the launch vehicle were then tested individually. For example, complete tanking tests were conducted in which the fuel and liquid oxygen tanks were loaded and pressurized to flight pressure. This test is performed to determine if any leaks are in the systems and also to check out the controls related to each system. During this test on the MA-6 launch vehicle no major leaks were evident; however, some minor leaks were discovered and subsequently corrected.

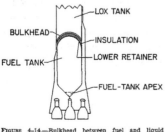

FIGURE 4-14.—Bulkhead between fuel and liquid oxygen tanks.

The autopilot system, as another example, is also tested; however, before autopilot tests are conducted on the launch complex the gyro packages are calibrated in a laboratory at Cape Canaveral with special testing equipment. These packages are also electrically mated to the abort sensing control package which is part of the ASIS previously discussed. During these laboratory and systems tests, various anomalies were uncovered in the gyro package and the ASIS control package. These packages were replaced and systems tests were completed satisfactorily on the launch vehicle. All launch-vehicle systems were then tested simultaneously in a test commonly known as the launch-vehicle flight acceptance composite test (FACT). This test is conducted to determine that all launch-vehicle systems are compatible so that each system will not adversely affect the operation of another. The launch vehicle FACT must be successfully accomplished before the spacecraft is electrically mated to the launch vehicle.

After electrical mating, the launch vehicle and spacecraft participate jointly in all tests. These tests are discussed in paper 5 on spacecraft preparation. The first attempt to launch MA-6 was on January 27, 1962. The launch vehicle was loaded with fuel on January 24. However, the mission was canceled because of excessive cloud cover in the launch area and was rescheduled for February 1; so the fuel tank was drained. On January 30, the fuel tank was again loaded; however, normal inspection procedures disclosed that the insulation-retaining bulkhead in the fuel tanks was leaking. Figure 4-14 shows a sketch of this bulkhead.

This partition is actually made up of three separate pieces. The top line represents the main bulkhead that provides structural integrity. Below this bulkhead is 1½ inches of insulating material and the lower line is the retaining bulkhead whose only purpose is to support the insulation. The leak was in the lower retainer and this had allowed fuel to soak into the insulation and become trapped. This trapped fuel, the amount of which was unknown at the time, could possibly have caused excessive inertia loads to be applied to the very thin retainer. After a careful study of the possible effects connected with this problem, it was decided that sufficient flight experience had been obtained on previous launch vehicles without the retainer to justify removing the retainer from this launch vehicle. Therefore, the sustainer engine was removed, the lower apex of the fuel tank was removed and a scaffold was built inside of the launch vehicle up to the retainer. After the retainer was removed and all systems were reconnected, a complete test program was rerun on every system disturbed by the modification. The simulated flight test was rerun on February 16, 1962, and the launch vehicle was again loaded with fuel in preparation for launch on February 20.

Countdown

During the actual launch countdown, two problems were experienced — one with the launch vehicle and one with ground support equipment on the complex. The first problem, the one involving a launch-vehicle system, occurred at T - 120 minutes and involved a malfunction of the guidance rate beacon. The beacon was replaced and checked out satisfactorily.

The other problem concerned the pumping system that loads liquid oxygen aboard the launch vehicle. This problem occurred during liquid oxygen tanking at T-22 minutes. The outlet valve in the main liquid oxygen pump failed in the closed position, but a smaller secondary pump was switched into the circuit to complete the tanking operation with no further incident. The remaining part of the countdown was performed smoothly and without trouble of any kind. At lift-off, blockhouse equipment indicated that all systems were functioning properly, and about 5 minutes later, the, report from the Mercury Control Center of successful insertion proved that the launch vehicle had performed its job nearly to perfection.

References

1. Program Office, Mercury/Atlas Launch Vehicle: Mercury/Atlas Launch Vehicle Factory Rollout Inspection General Procedures and Organization - Pilot Safety Program of the Atlas Launch Vehicle for NASA Project Mercury. Rep. no. TOR-594 (1101) RP-3 (Contract no. AF04 (647)-930), Aerospace Corp., El Segundo, Calif. Oct. 31, 1961.

2. Program Office, Mercury/Atlas Launch Vehicle: Mercury/Atlas Launch Vehicle Plight Safety Review General Operating Procedures and Organization — Pilot Safety Program of Atlas Launch Vehicle for NASA Project Mercury. Rep. no. TOR594 (1101) RP-4 (Contract no. AF04 (647) -930), Aerospace Corp., El Segundo, Calif., Oct. 31, 1961.

5. SPACECRAFT PREPARATION AND CHECKOUT

By G. MERRITT PRESTON, Chief, Preflight Operations Division, NASA Manned Spacecraft Center; and J. J. WILLIAMS, Preflight Operations Division, NASA Manned Spacecraft Center

Summary

Friendship 7 arrived at Cape Canaveral on August 27, 1961, and was launched February 20, 1962. The spacecraft underwent detailed system-by-system tests after arrival to verify its configuration. The configuration was changed as a result of information obtained from the MA-5 orbital flight. After the design changes were incorporated, the spacecraft underwent final hangar systems tests.

At the launch complex the spacecraft was mated to its launch vehicle and combined tests were conducted to ascertain the compatibility of the spacecraft, launch vehicle, and supporting range instrumentation. After ascertaining the compatibility and functional capabilities of all elements, the space vehicle was successfully launched February 20th.

Introduction

Friendship Seven arrived at Cape Canaveral, Fla., on August 27, 1961, for final preparation for flight. It was in checkout at Cape Canaveral for 166 working days. This appears to be a long time; therefore, the philosophy of operations for Project Mercury that dictates such a lengthy checkout at Cape Canaveral is discussed.

The reliability of the Mercury project was established by a step-by-step developmental flight program and a repeated detailed examination of the spacecraft and its systems.

Because of the urgency of the program, all spacecraft produced were used for flight testing and none were available for developmental testing in the laboratories until late in the program. Therefore, the preflight operations conducted at Cape Canaveral on the various spacecraft served not only to prepare that particular craft for flight, but it also was part of the design evaluation of the spacecraft that is typical of this type of program. The flight tests also contributed to this design evaluation.

The detailed examination of the spacecraft design, therefore, was primarily conducted at Cape Canaveral.

This examination involved functional testing of the spacecraft systems, observing in detail the performance of the systems. These tests were repeated often and duplicated as near as possible in different flight environments and modes. During these tests, all discrepancies, no matter how trivial, were scrutinized for their significance. Design changes indicated by these tests and the flight tests were incorporated as rapidly as possible so that the optimum spacecraft configuration was flown.

Astronauts Glenn and Carpenter participated in all system checkouts at Cape Canaveral and reviewed all design changes. This participation allowed for intimate familiarization with the spacecraft and a better understanding of its system.

Time Utilization

By examining the expenditure of the 166 days that Friendship Seven spent at Cape Canaveral, the effort spent in testing and modifying the spacecraft can be seen. The checkout of the spacecraft itself was conducted in Hangar S compound at Cape Canaveral, Florida, and lasted approximately 133 days as shown in figure 5-1. This checkout was followed by launch complex operations in which the launch vehicle and spacecraft were mated and testing was conducted to assure that the launch vehicle and spacecraft were mechanically, electrically, and radio-frequency compatible. This testing was followed by final preparations and assembly for the launch.

The time spent in the hangar can be initially broken into two parts:

(1) Work was performed on the spacecraft such as assembly and servicing the spacecraft and incorporating design changes that had been dictated by previous flights. This work took approximately 63 of the 133 days spent in the hangar.

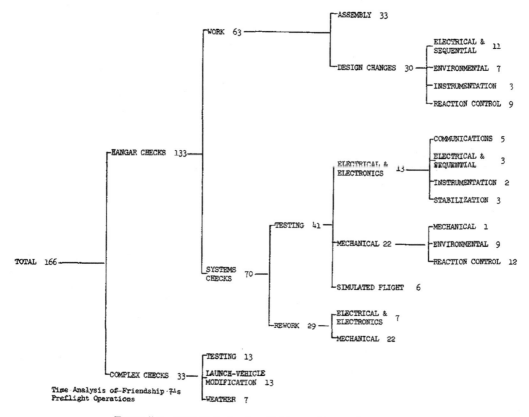

FIGURE 5-1.—Time analysis of preflight operations on Friendship 7.

(2) Systems testing, troubleshooting and replacing of components as the result of this testing were also conducted in the hangar. Seventy days were spent in this phase of the operation. The launch-complex operations required 33 days to perform.

A further breakdown of the aforementioned categories would show that 33 days were spent during the work periods to assemble and service the spacecraft in the hangar. Thirty days were spent incorporating design changes as the result of the previous flight. Of these 30 days, 11 were spent changing the wiring of the electrical and sequential system; 7 were spent modifying the environmental control system; 3 were spent in adding instrumentation; and 9 were spent modifying the reaction control system.

Of the 70 days spent in testing and taking corrective action, 41 were used for actual testing. Of these 41 days, 13 were used by the electrical and electronic systems, 22 for the mechanical systems, and 6 for overall simulated flight tests. Figure 5-1 shows a further breakdown of the electrical and electronic systems tests and the mechanical systems tests.

Twenty-nine days were spent troubleshooting and replacing components as the result of troubles encountered during these test periods. Of these 29 days, 7 days were spent on the electrical and electronic systems and 22 days were spent on the mechanical system. A majority of the troubles with the mechanical system were associated with the environmental control system which took 15 days to correct.

Of the 33 days spent on the launch pad (actually, 43 days were spent on the pad, however, 10 of these were spent troubleshooting and have been included in the above analysis) 13 days were spent in pad testing; 13 days were required to modify the launch vehicle as a result of a malfunction of the fueling system, and there were 7 days of weather delays.

Spacecraft Design Changes

Friendship 7 had several design changes made at Cape Canaveral as a result of information obtained from the orbital flight of spacecraft 9. A total of 255 changes were made in the spacecraft while at the Cape. Some of these items will be discussed subsequently for each of the major spacecraft systems.

Reaction Control System

The following changes were made to the reaction control system:

(1) Plastic flare seals were removed from the automatic system inlet and outlet connections to the thrust chamber solenoid valves and replaced with soft aluminum washers (fig. 5-2). Tests conducted at Cape Canaveral and McDonnell Aircraft Corp., St. Louis, Mo., revealed that the plastic seals resulted in relaxed torque on the tube fittings at ambient temperatures. It was also shown that this condition was further aggravated by the high temperatures found in the vicinity of the thrust chambers.

(2) Heat sinks were added to both manual and automatic roll thruster assemblies (fig. 5-3). Tests at McDonnell Aircraft Corp., St. Louis, Mo., and data obtained from the spacecraft 9 flight indicated a need for removing the excess heat generated by the thrusters. This heat resulted in undesirable fuel temperatures.

Electrical System

The following changes were made to the electrical system:

(1) The fuses were removed from the standby inverter circuit for the manual mode of operation.

FIGURE 5-2.—Reaction Control System flare seals.

(2) The fuse holders were structurally reinforced because of the record of mechanical failures that they have experienced in the past.

(3) Indicator lights were added to enable the pilot to determine which inverter was powering the bus.

(4) An auxiliary battery was added to the maximum altitude sensor wiring to eliminate voltage transients in the circuits.

(5) The abort signal to the sensor was interlocked with a spacecraft separation signal to prevent the tower from jettisoning prematurely.

FIGURE 5-3.—Heat sinks added to roll thruster assemblies in reaction control system.

(6) A camera programming system was added to enable the cameras to run at high and low speeds so that the film would last throughout the mission.

(7) Wiring to the thruster solenoids was brought through a common connector so that the solenoids could be disabled for long duration autopilot checkout to prevent electrical overheating of the solenoids during testing.

Environmental Control System

The following changes were made to the environmental control system:

(1) Advanced water-type heat exchangers (cold plates) were installed under the 150 and 250 main inverters (fig. 5-4). The results obtained on the spacecraft 9 flight indicated insufficient cooling.

FIGURE 5-4.—Inverter cooling plate.

FIGURE 5-5.—Cooling fan duct inlet screens for Friendship 7 and Spacecraft 9.

FIGURE 5-6.—Debris found in cooling fan of Spacecraft 9 after flight.

(2) The cooling fan duct inlet screens were changed to screens having 0.06-inch diameter holes. (See figure 5-5.) This change was required when the postflight inspection of spacecraft 9 revealed that the cabin fan was jammed by small bits of metal, fabric, and rubber, shown in figure 5-6, which were ingested into the fan under zero-g conditions through the 0.25-inch diameter holes in the inverter cooling duct screens.

(3) The aluminum check valves in the freon-water inverter cooling system were replaced with stainless steel valves. This change was made when it was found that the aluminum check valves tended to corrode and remained in either the open or closed position.

(4) An indicator was installed on the instrument panel to provide the astronaut with an indication of heat exchanger exhaust temperature in order to control the cooling flow in the suit circuit.

Automatic Stabilization and Control System

The following changes were made to the automatic stabilization and control system:

(1) The autopilot was replaced with a later model which contained new logic circuitry. This circuitry prevented erroneous orbit pulses due to intermittent sector switching. This change was designed to conserve reaction control system fuel.

(2) Fuses were added to the power leads to the rate gyros to prevent loss of an inverter in the event of a malfunction in the gyro.

(3) Heater blankets were added to the scanners. It was found that the effectiveness of the scanners was reduced unless the scanner bolometer was maintained at 75°.

Instrumentation System

The following changes were made to the instrumentation system:

(1) An instrument package was changed from a position near the right foot. This change was made to avoid interference with the pilot's right foot during flight and to facilitate removal of the hatch for egress.

(2) Temperature pickups were added to the 1-pound automatic and manual roll thrusters to monitor the effectiveness of the new heat sinks.

(3) A manual blood pressure measuring system was installed.

Miscellaneous Changes

Two other design changes were made. They are:

(1) A personal equipment container was manufactured and installed to provide a place for storing the items which the astronaut took on the flight to perform the activities required of him.

(2) Removable filters were installed on the pilot's observation window to provide protection to the astronaut when exposed to direct sunlight.

Hangar S Systems Tests

Checkout operations at Hangar S consisted of individual systems tests followed by a simulated flight test with all systems operating in a manner approaching flight conditions as nearly as possible. The following system tests were performed on Friendship 7:

(1) Electrical power
(2) Instrumentation
(3) Sequential
(4) Environmental control
(5) Communications
(6) Reaction control
(7) Communications radiation tower test
(8) Automatic stabilization and control
(9) Altitude chamber tests

The electrical power system test was the initial test performed on the spacecraft. The test determined if power could be safely applied to the control and power distribution system of the spacecraft. The test also checked automatic and manual a-c inverter switching. Figure 5-7 shows the spacecraft cabled up for this test. Similar cabling is used in many of the following tests. Power surges on the d-c bus were experienced when the 150 v-amp main inverter was switched on the line. Because an excessive number of power surges occurred, the inverter was replaced. Previous experience has indicated that such inverters often fail completely at a later date.

Figure 5-7.—Spacecraft cabled for electrical power system tests.

Figure 5-8.—Bench setup for testing spacecraft instrumentation.

The instrumentation system was thoroughly tested and calibrated on the bench as shown in figure 5-8 at Cape Canaveral prior to its installation in the spacecraft. The system was tested again after it had been installed in the spacecraft. The primary purpose of the latter test is fourfold: (1) to determine the error, if any, between the hardline signal normally used to transmit data to the blockhouse or other test equipment in the hangar and the signal radiated through the transmitter; (2) to determine possible system interference with the system operating in the spacecraft; (3) to make a single-point calibration check of the complete calibration made on the bench; and (4) to insure that the system was satisfactory to support other spacecraft tests. No major system discrepancies were uncovered during this testing.

The sequential system test provides for the checkout of the automatic and manual sequential system. The sequential system testing may be broken down into four major phases; namely, (1) launch sequence, (2) orbit sequence, (3) escape or abort sequence, and (4) recovery sequence. Control signals were fed to the various inputs of the sequential system and the output functions of the system were monitored. The maximum altitude sensor actuates the jettisoning of the escape tower. Although it is called a maximum-altitude sensor, it is really a variable timer whose time delay depends on the existence of an abort signal and the time from lift-off that the abort signal occurred. During the abort phase in the simulated flight of Friendship 7, it was determined that this sensor actuated early. The sensor was replaced and the abort runs were repeated successfully.

The environmental control system (ECS) checkout conducted prior to the altitude-chamber test determined the functional operation of the individual components of the ECS system. The oxygen bottles are serviced to operating pressure at this time for the altitude-chamber test.

As the result of testing, the following major ECS discrepancies were uncovered:

(1) Excessive leakage of the high-pressure oxygen shutoff valve was noted. The stem was removed from the valve body and the lubricant was found to have hardened on the O rings and sealing surfaces (fig. 5-9). The O rings and backup seals were replaced, and the component was lubricated and replaced in the valve body. The valve continued to show a leakage caused by the seals not seating properly. The leakage was reduced to within the specification volume by employing a specific procedure in opening the valve. The repeatability of this procedure was validated; that is, the leakage rate remained within specifications after following the opening procedure.

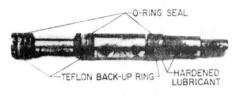

FIGURE 5-9.—High-pressure shutoff oxygen valve stem shown with hardened lubricant.

(2) The oxygen-flow-sensor warning light failed to come on when the primary bottle was exhausted. The sensors have shown erratic performance in the past. It was decided that the ground and flight monitoring was adequate information to warn the pilot of oxygen quantity.

(3) The high-pressure oxygen regulator showed an external leak. The valve body was found to be defective. The defect was repaired stopping the external leak.

(4) The aluminum freon check valve stuck open. These valves were replaced with newly designed stainless steel check valves which operated properly.

The primary purpose of the communication system tests was to determine the electrical characteristics of the individual components that comprise the onboard communications system. Although the test equipment was physically located approximately 100 feet away in the checkout trailer (fig. 5-10), the test was equivalent to a bench test with the components in the spacecraft. The test revealed the following discrepancies:

During HF rescue voice receiver checks, the HF output change was 14 decibels with an input signal level change from 5 to 50,000 microvolts. The specification requirement was for not more than 10 decibels. With an input variation of from 10 to 50

FIGURE 5-10.—Communications system test equipment in checkout trailer.

microvolts, the output change was 7 decibels. This deviation from specification was not considered serious and was considered acceptable. It is well to point out at this time that detailed specifications have been established for the performance of spacecraft equipment. However, during spacecraft checkout, when the performance does not meet these specifications, the required performance is reviewed by considering the requirements for the particular flight and the increase in knowledge about the performance requirements that has been obtained since they were originally established. Thus, equipment is not arbitrarily changed when it does not meet specification.

FIGURE 5–11.—Spacecraft in reaction control system facility for functional test of system.

FIGURE 5–12.—Spacecraft being installed on communications tower for communication antenna tests.

The reaction control system (RCS) checkout procedure determined the condition and operation of the RCS system. Tests were conducted in a special test cell (fig. 5-11). Several gas checks were employed to determine the overall system gas integrity. The system was then filled with 35-percent hydrogen peroxide (H_2O_2) for 24 hours to monitor decomposition pressure rise and to determine system cleanliness as well as to precondition the system for use with 90-percent H_2O_2. Following the 35 percent H_2O_2 surveillance, a hydrostatic check of each system using 35-percent H_2O_2 was made to determine system liquid integrity. A 24-hour surveillance using 90-percent H_2O_2 was then conducted. This surveillance was followed by a functional check that pressurized the system and static firing each of the thrusters.

The system performed very well during these tests, except the manual proportional control mode had slightly higher stick forces than desired. These forces, however, were acceptable to Astronauts Glenn and Carpenter.

Communications system tests were made on the radiation tower to determine the HF characteristics of the bicone antenna. Tests of the other communications system components were also made at the same time to permit the Atlantic Missile Range to evaluate the system. For this test the spacecraft was placed on a 44-foot-high wooden tower as shown in figure 5-12. The test is conducted with no ground servicing equipment (GSE) cables connected to the spacecraft; thus the flight configuration was simulated as closely as possible.

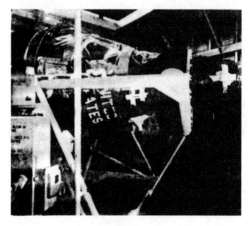

FIGURE 5–13.—Dynamic tests of autopilot in dynamic fixture.

During the test it was noted that the auxiliary UHF beacon caused some interference on the UHF voice transmission. Since the interference was not great and did not affect the intelligibility of UHF voice transmissions, system operation was considered satisfactory for flight.

The automatic stabilization and control system (ASCS) checkout of the spacecraft was divided into two parts: a static test and a dynamic test. The dynamic test was conducted with the spacecraft in a dynamic fixture (fig. 5-13) which could be rotated at constant rates of roll and pitch. Yaw dynamic tests were conducted by rolling the spacecraft 90° and pitching. The logic of the system was tested by using an automatic tester located in the checkout trailer as shown in figure 5-14.

The autopilot passed the static test with only minor discrepancies. However, the dynamic portion of the test was terminated with a failure of the yaw repeater loop. A shorted capacitor was discovered in the repeater motor circuitry. A new autopilot passed both the static and dynamic test with no system discrepancies.

The altitude-chamber tests (fig. 5-15) were used to determine the operating characteristics of the overall environmental control system (ECS). The astronaut was suited and connected into the ECS for the first time during the altitude-chamber tests. The chamber was pumped down to a simulated altitude of approximately 125,000 feet and a simulated mission was conducted.

FIGURE 5-16.—Plugged inlet part of cooling plate for 150 v-amp inverter.

FIGURE 5-14.—Automatic checkout equipment for checking logic of autopilot.

FIGURE 5-17.—Spacecraft during final simulated flight test in Hangar S.

FIGURE 5-15.—Overall evaluation of the environmental control system in altitude chamber.

Test data indicated that the 150 v-amp inverter overheated during the first three runs. The flow orifices were increased in size for the fourth run.

During the fourth run, the 150 v-amp inverter still overheated while the 250 v-amp inverter remained cool. A subsequent investigation indicated that the flow passages in the water-type heat exchanger (cooling plate) were plugged. It was found that sealant used in the construction of the cooling plates had penetrated into the flow passages as shown in figure 5-16. The passages were cleaned and the original orifices were reinstalled. Flow tests on the final installation showed that the system was clean.

Friendship 7 began its hangar simulated flight test on November 25, 1961. This test was completed on December 12, 1961. When the launch was rescheduled for January 1962, a rerun of the hangar simulated flight was made. The second simulated flight test series began December 19, 1961, and were successfully completed on December 21, 1961. At this time, the spacecraft was considered functionally ready for launch pad operations.

This simulated flight test was designed to accomplish the following test objectives:

(1) To reverify proper operation of individual systems

(2) To insure proper operation of the sequence system through all modes, including abort and emergency override

(3) To demonstrate intrasystem compatibility when all systems were operating concurrently

(4) To verify proper operation of spacecraft systems when the flight conditions are simulated as nearly as practicable

The configuration of the spacecraft during the simulated flight test was as follows and is shown in figure 5-17.

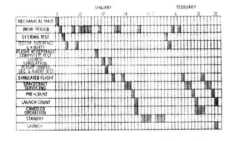

FIGURE 5-18.—Launch complex testing of Friendship 7.

FIGURE 5-19.—Spacecraft on Atlas launch vehicle undergoing launch complex testing.

(1) The spacecraft was installed on the launch-vehicle adapter, and was electrically connected to the adapter.

(2) The escape tower without the escape rockets was installed on the spacecraft.

(3) An absolute minimum of GSE cabling was connected to the spacecraft to simulate actual flight configuration as closely as possible.

(4) Two-tenths ampere fuses, used to simulate squibs, were installed at actual squib locations.

(5) All squib firing-circuit wiring was in flight configuration.

(6) The astronaut was suited and in the spacecraft for the systems test.

(7) Recorders were connected to monitor all squib circuits to verify proper firing times.

Malfunctions during the simulated flight test were as follows:

(1) The no. 2 suit fan showed intermittent flow characteristics. Further troubleshooting led to replacement of no. 2 suit fan, no. 1 and no. 2 check valves, and the negative pressure relief valve.

(2) The C-band beacon operated intermittently at low voltage. The beacon was replaced.

(3) The instrumentation time correlation clock was replaced because of an inoperative time-zero light.

(4) Plastic compound was found on a fuse block contact. The compound was removed and all fuse blocks were checked for the compound, but none was found.

(5) The satellite clock jammed during mechanical reset. It was replaced.

(6) The suit-fan toggle switch could be made to break contact when a slight force was placed on the toggle while it was in the no. 2 position. The switch was replaced.

(7) The microphone wires in the astronaut's helmet were found reversed, and the helmet was repaired.

(8) The emergency 10-second retrorocket fire relay timed out in 5 seconds. The relay panel was replaced and reverified.

Launch Pad Operations

As originally planned the launch pad operations were scheduled for completion in 13 days; however, due to delays caused by changes in the spacecraft and launch vehicle after mating, trouble encountered during tests, and adverse weather conditions, the actual time the spacecraft spent on the launch pad is somewhat longer than the 13 days as shown in figure 5-18. For Friendship 7 this period extended from January 3, 1962, through February 20, 1962. Figure 5-19 shows the spacecraft on the launch vehicle during these launch pad operations.

FIGURE 5-20.—Automatic wiring checker used to check cabling involved in preflight operation.

The launch pad operations are listed below in the order in which they are normally performed:

(1) Launch complex checkout
(2) Interface inspection
(3) Mechanical mating
(4) Spacecraft systems test
(5) Electrical interface and aborts
(6) Flight acceptance composite tests
(7) Flight configuration sequence and aborts
(8) Launch simulation including reaction control thruster firings
(9) Simulated flight
(10) Pyrotechnic check
(11) Spacecraft servicing
(12) Precount
(13) Count

Each of these launch pad operations, together with the troubles encountered, subsequent delays, and repairs made, is discussed for the tests made during the checkout of the complex, the spacecraft, and the spacecraft-launch-vehicle combination.

Launch Complex Checkout

The launch complex checkout is performed to validate all the complex wiring and launch pad modifications prior to mating the spacecraft with the launch vehicle. An automatic wire checker is used in this operation. It is shown in figure 5-20.

There were no serious discrepancies discovered during this test and the complex was declared ready for receiving the spacecraft.

FIGURE 5-21.—View of spacecraft adapter being mounted on launch vehicle.

Interface Inspection

Prior to mating the spacecraft to the launch vehicle adapter, it was necessary to verify that all items in the adapter section had the proper clearance and that there was no debris. Figure 5-21 shows the adapter being mounted on the launch vehicle. Following the inspection and mating of the spacecraft to the launch vehicle, all openings into this vital section were sealed. These seals are not removed again except for emergency work and just prior to launch. The adapter and interface area met all requirements and the spacecraft was then mated with the launch vehicle.

FIGURE 5-23.—Blockhouse consoles for monitoring spacecraft operation during the launch complex operation.

FIGURE 5-22.—Mechanical mating of spacecraft to adapter.

Mechanical Mating

Mechanical mating was primarily a mechanical fit operation in which the spacecraft was placed on the launch-vehicle adapter (fig. 5-22) and the main clamp ring was installed. In addition, prior to hoisting the spacecraft to the top of the launch vehicle the retropackage pyrotechnics are checked for continuity and resistance and then shorted with shorting plugs. No unforeseen difficulties were experienced in mating spacecraft 13 to the launch vehicle; thus the MA-6 vehicle was ready for launch pad checkout.

FIGURE 5-24.—Emergency egress tower ramps in place during emergency egress tests.

Spacecraft Systems Tests

Spacecraft systems tests were performed primarily to check the spacecraft systems functionally and to check the remote-control capability from the blockhouse. Figure 5-23 shows the spacecraft blockhouse consoles.

Following this test the spacecraft was ready to proceed with tests which integrate the launch vehicle and the spacecraft. During these tests trouble was experienced with the coolant quantity indicating system indicator, the auxiliary beacon, and minor discrepancies in ground-servicing equipment (GSE) cabling. These items were corrected and those parts of this system tests required were rerun to validate the changes.

Electrical Interface and Aborts

The electrical interface and aborts test validated the spacecraft-launch-vehicle interface compatibility as well as redundant electrical and radio-frequency paths. For this test, GSE cables were installed. By use of the GSE cables this test also provides a means of checking the redundant paths used for abort at various times during the prelaunch and launch modes. This test was run with only minor discrepancies which were corrected, and then those portions of the test which had contained the discrepancies were rerun to validate the changes. However, during preparations for this test faulty automatic roll thrusters were discovered and these were replaced. At the same time, heat sinks were added to the thruster assemblies as a result of a design change. The silver batteries for the auxiliary beacons were also replaced. All of this special work resulted in a 4-day delay in testing.

Flight Acceptance Composite Test

The flight acceptance composite test was designed to prove compatibility of the spacecraft radio-frequency systems with the Atlantic Missile Range and the launch vehicle and to prove that spacecraft and launch vehicle systems do not generate radio-frequency interference. This test was also used to verify satisfactory operation of spacecraft radio frequency with GSE cables installed. This test was completed with only a minor discrepancy which required a change of a freon check valve.

At this point, because of a projected change in launch schedule, the order of testing was changed and the launch simulation was scheduled next.

Launch Simulation

The launch simulation test was designed to validate the spacecraft and launch-vehicle systems in a launch configuration and to evaluate the launch-day procedures. At the same time this test provides an opportunity to practice emergency egress procedures (fig. 5-24). No problems were encountered in this test.

Flight Configuration Sequence and Aborts

The flight configuration sequence and aborts test provides for compatibility tests of spacecraft systems and launch-vehicle autopilot. The test also further checks the abort modes with the spacecraft in a flight configuration simulating launch and flight. This test was completed with no difficulties. The launch simulation test was repeated for practice for the launch crew.

Flight Safety Review

A flight safety review board was convened approximately five days before launch to review the spacecraft history. This review was conducted by the operations directors. The review board was composed of representatives of the Mercury Project Office, the Flight Safety Office, the astronauts, and the Preflight Operations Division. This board reviewed the status of all systems, approving all deviation from specifications found during spacecraft checkout. It was the responsibility of this board to commit the spacecraft to launch.

Simulated Flight

The simulated flight test completely checks all spacecraft systems during a simulated flight from liftoff to landing. The test also permits a check of the automatic launch-vehicle abort system at lift-off + 200 seconds.

During the simulated flight test on X-4 days for the attempted launch on January 23, 1962, it was noted that the oxygen consumption rate, with Astronaut Glenn in the system, was high intermittently. Checks the previous day with Astronaut Carpenter in the circuit had indicated normal consumption rates.

Friendship 7's environmental control system was different than any of those previously flown in that it had an orifice in the oxygen supply that provided a constant flow of about 1,000 cc/min. This flow was adequate for all normal astronaut demands. Originally, the environmental system contained a demand regulator which sensed a drop in pressure in the suit circuit indicating a need for oxygen. With the drop in pressure, the regulator would supply oxygen to bring the system up to operating pressure. With the addition of the constant bleed orifice it was not expected that the demand regulator would function often. Troubleshooting following the simulated flight indicated that one of the diaphragms in the demand regulator had developed a small leak. Since it was very difficult to remove the regulator because of its location, extensive testing was made to be sure that the trouble truly was in the regulator. Finally the decision was made to remove the regulator and it was found that the problem was not in the regulator but in the plumbing to the regulator. Here, a leaky joint was found.

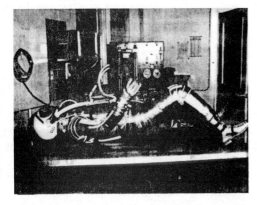

FIGURE 5-25.—Pressure tests of astronaut's pressure suit at 5 psi.

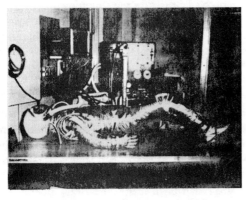

FIGURE 5-26.—Low pressure tests of suit.

WRIST CUFF SEAL

FIGURE 5-27.—Pressure-suit gloves showing cuff slip joint.

These incidents pointed out two principles that should be observed in design: First, inaccessibility of the demand regulator made it difficult to check the regulator adequately and caused delays at a crucial time in launch preparations (X-4 days). Secondly, inadequate test points made it impossible to diagnose the problem properly.

After the regulator was replaced and the plumbing repaired, another check was made with the astronaut in the suit circuit. These checks indicated some improvement in the oxygen consumption rate, but it was still much higher than it should have been.

After a detailed analysis of the system it was concluded that there must be a leak in the astronaut's pressure suit. Prior to any test involving the suit, it was pressure tested at 5 psi (fig. 5-25). However, the suit, during normal flight, operates at a pressure of only a few inches of water above cabin pressure and would operate at 5 Psi differential only if the cabin should become decompressed. No leak checks previously had ever been made at these low pressures. Inspection of the suit cuff and zipper seals indicated that pressure was required to effect a good seal at these points. Upon testing of the suit at low pressure (fig. 5-26) it was found that the cuff seals at the wrists (fig. 5-27) leaked as the cuff was rotated.

This particular experience leads to some additional conclusions:

(1) The necessity for functional testing at the launch site is proven again.

(2) The necessity for integrated system testing is demonstrated.

(3) The fallacy of assuming that one test condition is equivalent to another is pointed out. Testing should be conducted under conditions as near to all flight conditions as possible.

Upon checking the system after the previously mentioned faults were corrected, it was found that the system used oxygen at approximately the 1,000 cc/min rate provided by the constant bleed orifice.

Since the constant bleed orifice was added to an already automatic system, this bleed changed the pressure levels in a delicately balanced system. Without the bleed the small leaks in the suit circuit would have been inconsequential. However, with the bleed, the suit leakage was causing the demand regulator to flow extra oxygen into the suit, and thus the oxygen consumption was increased. This seemingly small change, therefore, produced large effects on the performance of the system. It must be pointed out that the addition of the constant bleed did provide additional safety by reducing carbon dioxide concentration in the suit circuit and provided positive oxygen flow regardless of the regulator operation.

Of additional interest was the fact that none of these discrepancies were noted during the altitude chamber tests even though the same test procedures were used. Apparently, the position of the cuff seals during this test prevented them from leaking. Also, the plumbing leak in the demand regulator plumbing must have occurred subsequent to the chamber tests. This indicates the desirability of repeating tests of systems at discrete times.

A simulated flight was repeated to verify the changes in the environmental control system.

Spacecraft Servicing

Various items of work were then performed to put the spacecraft in a flight-ready condition. These items are filling the oxygen bottles, servicing and installing the onboard tape recorders, servicing the landing-bag release system, and many other items of a purely mechanical nature. This work was accomplished in the allotted time and the precount was scheduled.

Precount and Launch Count

Precount — The countdown is performed in two parts. The first part, known as the precount, is primarily a check of the various spacecraft systems. Following completion of this first part of the countdown, there is an approximate 15-hour hold for pyrotechnic check, electrical connection, and peroxide system servicing and surveillance. Both the precount and the 15 hour hold operations were performed without discrepancies, and the final part, the launch count, was started. This count proceeded to T-13 minutes at which time the launch was canceled for the day because of adverse weather conditions.

Difficulties encountered after cancellation. — After the cancellation at T-13 minutes on January 27, 1962, it was decided to replace the carbon dioxide absorber unit because it had approached the end of its service life. The peroxide system also had to be drained and flushed to prevent corrosion, and the pyrotechnics were disconnected and shorted as a safety measure. This work was accomplished in 1 day. At this time, some of the launch-vehicle systems were being revalidated. During the tanking test of the launch vehicle, a leak was discovered in the inner bulkhead of the fuel tank and required 4 to 6 days to repair. The repairing of this leak, the necessary retesting and launch preparation after the repair had been made, and other operational considerations dictated rescheduling of the launch for February 13, 1962. The launch date was rescheduled to February 14 sometime later because all of the aforementioned work had not been completed. During this delay, all six flight batteries and the parachutes were replaced. Portions of the normal complex testing were rerun to verify launch status. The precount was started again on February 13, 15, and 16, but it was canceled each time because of adverse weather conditions. The launch was then rescheduled for February 20, 1962.

Launch count. — During the launch count on February 20 all systems were energized and final overall checks were made. The count started at T-390 minutes by installing and connecting the escape rocket igniter. The service structure was then cleared and the spacecraft was powered to verify no inadvertent pyrotechnic ignition. The personnel then returned to the service structure to prepare for static firing of the reaction control system at T-250 minutes. Following the reaction control static firing the spacecraft was then prepared for astronaut boarding at T-120 minutes. The hatch was put in place at T-90 minutes. During installation a bolt was broken, and the hatch had to be removed to replace the bolt. From T-90 to T-55 final mechanical work and spacecraft checks were made and the service was evacuated and moved away from the launch vehicle. At approximately T-35 minutes filling of the liquid oxygen tanks began and final spacecraft and launch vehicle systems checks were started. At T-10 minutes the spacecraft went on internal power, and the launch vehicle went to internal power at T-3 minutes. At T-35 seconds the spacecraft umbilical was ejected and at T-0 the main engines started. At about T + 4 seconds lift-off occurred and the flight was underway as shown in figure 5-28.

Concluding Remarks

In conclusion, the flight success of Mercury has in part been achieved by: (1) repeated testing to uncover systems weakness; (2) particular attention to details that might lead to mission failure, (3) integrated flight simulations that assured compatibility among the spacecraft systems and among the spacecraft, launch vehicle, and range, and (4) a continual updating of the spacecraft configuration taking full advantage of previous flight experience. This policy has led to lengthy checkout periods at Cape Canaveral; however, a single flight mission failure would have caused even longer delays in the overall Mercury program.

FIGURE 5-28.—Lift-off of MA-6 from launch complex
at Cape Canaveral, Fla.

6. FLIGHT CONTROL AND FLIGHT PLAN

By CHRISTOPHER C. KRAFT, Jr., Chief, Flight Operations Division, NASA Manned Spacecraft Center; TECWYN ROBERTS, Flight Operations Division, NASA Manned Spacecraft Center; EUGENE F. KRANZ, Flight Operations Division, NASA Manned Spacecraft Center; and C. FREDERICK MATTHEWS, Flight Operations Division, NASA Manned Spacecraft Center

Summary

A number of malfunctions occurred during this flight which caused some concern to the flight control team. These included the malfunction of the automatic control system, and what later proved to be the false indication of heat-shield deployment. However, the presence of the astronaut onboard the spacecraft made these malfunctions of a minor nature. The astronaut's ability to evaluate the performance of the spacecraft systems and take corrective action, and his excellent method of reporting these results to the ground, resulted in the successful completion of the MA-6 flight.

Introduction

It is the intent of this report to give a brief outline of the flight plan of the MA-6 flight and primarily the procedure used to perform flight control. In addition, some of the pertinent flight test results will be given.

Flight Plan

A detailed outline of the flight plan is given as follows:

The astronaut was to evaluate the various modes of control available in the spacecraft and to report on the capabilities of these various systems. He was to determine his visual reference capabilities, that is, his ability to determine attitude by observing the horizon and/or the stars by using the window and periscope and his ability to obtain this reference on both the light and dark sides of the earth. Certain specific maneuvers were set up to provide information on this capability. The effects of weightlessness for extended periods were to be determined by his ability to perform in this environment and, here again, specific tests were set up to aid in this determination. The astronaut was to perform management of onboard systems such as cabin and suit cooling, use of a-c and d-c power, and so forth, and he was to report on the performance of all of the spacecraft systems. The astronaut was to determine his ability to navigate by both earth and star reference. He was to perform visual observations of astronomical and scientific interests, including weather observations. Finally, he was to report on any unusual phenomena within or outside of the spacecraft. Although the malfunction of some of the spacecraft systems may have altered the flight plan to some extent, it is felt that the flight test results achieved a great majority of the objectives laid down and that generally speaking the flight test was extremely successful.

The following description of the mission is given in chronological order so that appreciation of the flight control problems can be understood, and in this manner the flight tests results are given.

Countdown

The countdown for launching the Mercury - Atlas vehicle is conducted in two parts. The first part is conducted on the day before the launch and lasts approximately 4 hours. During this period detailed tests of all of the spacecraft systems are performed and those interface connections important to these systems are verified. This part of the countdown was conducted with no major problems or holds resulting. Approximately 17½ hours separated the end of this count and the beginning of the final countdown, and during this period the spacecraft pyrotechnics were installed and connected and certain expendables such as fuel and oxygen were loaded.

At T-390 minutes the countdown was resumed and progressed without any unusual instance until T -120 minutes. During this period additional spacecraft systems checkouts were performed and the major portion of the launch-vehicle countdown was begun. At T-120 minutes a built-in hold of 90 minutes had been scheduled to assure that all systems had been given sufficient time for checkout before astronaut insertion. During this period a problem developed with the guidance system rate beacon in the launch vehicle causing an additional 45 minute hold, and an additional 10 minutes were required to repair a broken microphone bracket in the astronaut's helmet after the astronaut insertion procedure had been started. The countdown proceeded to T-60 minutes when a 40 minute hold was required to replace a broken bolt because of misalignment on the spacecraft's hatch attachment. At T -45 minutes, a 15 minute hold was required to add fuel to the launch vehicle; and at T-22 minutes an additional 25 minutes was required for filling the liquid-oxygen tanks as a result of a minor malfunction in the ground support equipment used to pump liquid oxygen into the launch vehicle. At T-6 minutes and 30 seconds, a 2 minute hold was required to make a quick check of the network computer at Bermuda. In general, the countdown was very smooth and extremely well executed. A feeling of confidence was noted in all concerned, including the astronaut, and it is probably more than significant that this feeling has existed on the last three Mercury-Atlas launches.

Powered Flight

The launch occurred at 9:47:39 a.m. e.s.t. on February 20, 1962. The powered portion of the flight which lasted 5 minutes and 1 second was completely normal and the astronaut was able to make all of the planned communications and observations throughout this period. Throughout this portion of the flight no abnormalities were noted in either the spacecraft systems or in the astronaut's physical condition. The launch-vehicle guidance system performed almost perfectly, and 10 seconds after cutoff the computer gave a "go" recommendation. The cutoff conditions obtained were excellent.

Table 6-1 presents the actual cutoff conditions that were obtained. A comparison of the planned and actual times at which the major events occurred are given in table 6-II and the times at which all of the network sites acquired and lost contact with the spacecraft are presented in table 6-III.

It might be noted that the flight test experience which had been achieved on the previous Mercury-Atlas orbital flights, that is, the MA-4 and MA-5, had given the flight control team an excellent opportunity to exercise control over the mission. These flights were, of course, much more difficult to control and complete successfully because of the lack of an astronaut within the spacecraft. All of the analyses and decisions had to be made on the basis of telemetered information from the spacecraft. The presence of an astronaut made the flight test much more simple to complete, primarily on the basis of astronaut observations and his capability of systems management. A manned flight, however, makes the job of monitoring spacecraft performance more complex because of the large number of backup and alternate systems from which the astronaut could choose.

TABLE 6-1 — Planned and Actual Flight Conditions	
Cutoff conditions:	
Altitude, ft	528,381
Velocity, ft/sec	25,730
Flight-path angle, deg	-0.047
Orbit parameters:	
Perigee altitude, nautical miles	86.92
Apogee altitude, nautical miles	140.92
Period, min: sec	88:29
Inclination angle, deg	32.54
Maximum conditions:	
Exit acceleration, g units	7.7
Exit dynamic pressure, lb/sq ft	982
Entry acceleration, g units	7.7
Entry dynamic pressure, lb/sq ft	472

Mission Rules

Previous to all of the flights, mission rules for all phases of the operation were established beginning with the countdown and ending with the recovery. The development of these rules was started a considerable length of time before any of the Mercury flight operations, and began to develop at the same time as the flight control concepts. The mission rules were established in an effort to take into account every conceivable situation which could occur onboard the spacecraft; that is, consideration was given to both the astronaut and the spacecraft systems, and to all of the conceivable ground equipment failures which could have a direct bearing on the flight operation. In addition, rules were established in an effort to handle a large number of launch vehicle malfunctions. These, of course, dealt primarily with the effects of a sudden cutoff condition and its effect on the spacecraft flight thereafter. These rules were established for the prelaunch, powered flight, and orbital flight phases of the mission.

TABLE 6-II. Sequence of Events During MA-6 Flight (All times are Eastern Standard)

Event	Planned time (a) hr:min:sec	Actual time, hr:min:sec
Booster-engine cutoff	00:02:11.4	00:02:09.6
Tower release	00:02:34.2	00:02:33.3
Escape rocket firing	00:02:34.2	00:02:33.4
Sustainer-engine cutoff (SECO)		00:05:01.4
Tail-off complete	00:05:03.8	00:05:02
Spacecraft separation	00:05:03.0	00.05.03.6
Retrofire initiation	04:32:58	04:33:08
Retro (left) No. 1	04:32:58	04:33:08
Retro (bottom) No. 2	04:33:03	04:33:13
Retro (right) No. 3	04:33:08	04:33:18
Retro assembly jettison	04:33:58	(b)
0.05g relay	04:43:53	04:43:31(c)
Drogue parachute deployment	04:50:00	04:49:17.2
Main parachute deployment	04:50:36	04:50:11
Main parachute jettison (water impact)	04:55:22	01:55:23

a) Preflight calculated, based on nominal Atlas performance.
b) Retro assembly kept on during reentry.
c) The 0.05g relay was actuated manually by astronaut when he was in a "small g field."

Because of the complexity of the entire operation and the critical time element of powered flight, it was felt and borne out by flight experience that such a set of rules were an absolute necessity. Of course, it is impossible to think of everything that can happen but if most of the contingencies have been anticipated along with the procedures to handle these situations, the time available can be used to concentrate on the unexpected. The occurrence of the heat shield deploy signal in this flight is an example of one of these unforeseen circumstances.

Flight Test Results

The rest of this paper deals primarily with the flight test results and flight control problems which developed throughout the three orbit mission. The observations made by the astronaut and his evaluation of the mission are presented in paper 12 by Astronaut John H. Glenn, Jr.

After separation of the spacecraft from the launch vehicle, the astronaut was given all the pertinent data involved with orbit parameters and the retrofire times necessary had immediate reentry been required. Following these transmissions, which were primarily from the Bermuda site, the astronaut made the planned checks of all of the spacecraft control modes using both the automatic and manual proportional systems. All of these checks indicated that all of the control systems were operating satisfactorily. Also, the astronaut reported that he felt no ill effects as a result of going from high accelerations to weightlessness, that he felt he was in excellent condition and, as the two previous astronauts had commented, that he was greatly impressed with the view from this altitude.

The first orbit went exactly as planned and both the astronaut and the spacecraft performed perfectly. Over the Canary Islands' site, the astronaut's air-to-ground transmissions were patched to the voice network and in turn to the Mercury Control Center and provided the control center and all other voice sites the capability of monitoring the transmissions to and from the spacecraft in real time. This condition existed throughout all three orbits from all sites having voice to the control center and provided the best tool for maintaining surveillance of the flight. (See appendix A.)

TABLE 6-III.— *Network Acquisition Times for MA-6 Flight*				
Station	Telemetry signal duration, Hr:min:sec		Voice reception	
	Acquisition	Loss	Frequency	Duration, hr:min:sec
Canaveral	00:00:00	00:06:20	UHF	00:00:00 to 00:05:30
Bermuda	00:03:02	00:10:26	UHF	00:03:30 to 00:09:30
			HF	00:11:00 to few sec
Canary Islands	00:14:15	00:21:23	UHF	00:15:00 to 00:23:00
			HF	00:13:00 to 00:14:00
Atlantic Ship	Not in range			
Kano	00:21:13	00:28:21	UHF	00:22:00 to 00:29:00
Zanzibar	00:29:51	00:37:51	UHF	00:30:00 to 00:38:00
Indian Ocean Ship	00:40:02	00:48:31	UHF	00:41:00 to 00:48:00
Muchea	00:49:21	00:57:55	UHF	00:50:00 to 00:58:30
Woomera	00:54:00	01:02:41	UHF	00:56:00 to 01:03:00
Canton	01:09:19	01:17:42	UHF	01:09:00 to 01:15:30
Hawaii	Not in range			
California	01:26:41	01:31:23	UHF	01:27:30 to 01:30:00
			HF	01:19:00 to 01:25:30
Guaymas	01:26:47	01:33:25	UHF	01:26:00 to 01:33:30
			HF	01:20:30 to 01:26:00
Texas	01:29:24	01:36:18	UHF	01:33:30 to 01:39:00
			HF	01:28:30 to 01:56:30
Eglin	01:32:00	01:37:05		
Canaveral	01:33:20	01:40:03	UHF	01:33:30 to 01:40:00
			HF	01:28:00 to 01:43:00
Bermuda	01:36:38	01:43:53	UHF	01:33:30 to 01:42:00
			HF	01:43:00 to 01:49:30
Canary Islands	01:47:55	01:53:58	UHF	01:49:00 to 01:55:00
Atlantic Ship	01:51:54	01:58:31	UHF	01:54:00 to 01:58:00
Kano	01:54:47	02:01:21	UHF	01:58:00 to 02:02:00
Zanzibar	02:04:05	02:10:51	UHF	02:04:00 to 02:11:00
Indian Ocean Ship	02:12:17	02:22:09	UHF	02:14:00 to 02:23:00
Muchea	02:22:51	02:31:23	UHF	02:25:00 to 02:32:30
Woomera	02:27:36	02:35:45	UHF	02:28:00 to 02:37:00
Canton	02:42:51	02:49:45	UHF	02:41:30 to 02:49:00
Hawaii	02:49:01	02:55:19	UHF	02:49:00 to 02:55:30
California	02:58:11	03:04:48	UHF	02:58:30 to 03:04:30
Guaymas	02:59:59	03:06:44	UHF	03:00:30 to 03:04:20
Texas	03:03:14	03:09:39	UHF	03:03:30 to 03:10:30
			HF	03:03:30 to 03:10:30
Eglin	03:05:35	03:12:07	UHF	03:07:30 to 03:12:30
Canaveral	03:06:51	03:13:46	UHF	03:07:00 to 03:14:00
			HF	03:05:00 to 03:13:30
Bermuda	03:09:56	03:17:03	UHF	03:00:30 to 03:16:30
			HF	03:17:00 to 03:18:00
Canary Islands	Not in range		HF	03:21:00 to 03:25:00
Atlantic Ship	03:24:44	03:32:25	UHF	03:26:00 to 03:34:00
			HF	03:26:00 to 03:27:30
Kano	Not in range			
Zanzibar	Not in range		HF	03:41:00 to 03:42:00
Indian Ocean Ship	03:46:55	03:56:49	HF	03:48:00 to 03:56:00
Muchea	03:56:31	04:04:12	UHF	03:58:30 to 04:04:00
			HF	03:57:00 to 03:58:00
Woomera	04:03:16	04:06:19	UHF	04:04:00 to 04:07:00
Canton	Not in range		HF	04:15:30 to 04:21:00
Hawaii	04:21:49	04:28:49	UHF	04:19:00 to 04:40:30
			HF	04:20:00 to 04:20:15
California	04:31:17	04:37:57	UHF	04:31:30 to 04:38:30
			HF	04:19:00 to 04:19:15
Guaymas	04:33:44	04:39:49	UHF	04:34:30 to 04:40:30
Texas	04:36:53	04:42:32	UHF	04:38:00 to 04:39:00
			UHF	04:41:00 to 04:43:00
			HF	04:36:00 to 04:45:00
Eglin	04:39:00	04:42:52	UHF	04:39:00 to 04:43:30
Canaveral	04:40:52	04:42:55	UHF	04:40:30 to 04:44:30
			HF	04:33:30 to 04:42:30
Canton			HF	04:35:15 to 04:35:30

Except for the control systems checks which were made periodically, the astronaut remained on the automatic system with brief periods on the fly-by-wire system which utilizes the automatic control jets. This procedure was planned so that a fixed attitude would be provided for radar tracking and so that the astronaut could make the necessary reports and observations during the first orbit. During the first orbit, it was obvious from the astronaut's reports that he could establish the pitch and yaw attitude of the spacecraft with precision by using the horizon on both the light and dark sides of the earth, and that he could also achieve a reasonable yaw reference. Aside from the xylose tablet taken over Kano, he had his first and only food (tube of applesauce) over Canton Island during this orbit and reported no problems with eating nor any noticeable discomforts following the intake of this food.

During the first orbit, the network radar systems were able to obtain excellent tracking data and this data, together with the data obtained at cutoff, provided very accurate information on the spacecraft position and orbit. As an example, between the time the spacecraft was inserted into orbit and the data were received from the Australian sites, the retrosequence times changed a total of only 7 seconds for retrofire at the end of 3 orbits. This indicated the accuracy of the orbit parameters. From this point to the end of the three orbit flight, using all of the available radar data, these times changed only 2 seconds. The final retrosequence time was 04:32:38 as compared with the time initially computed at cutoff of 04:32:47 and the time initially set into the clock on the ground before liftoff of 04:32:28. All of the network sites received data from the spacecraft and maintained communications with the astronaut from horizon to horizon, and everything progressed in a completely normal fashion. Because of the excellent condition of the astronaut and the spacecraft, there was no question about continuing into the second orbit, and a "go" decision was made among personnel at Guaymas, Mexico and the Mercury Control Center and "foremost" the astronaut himself.

Shortly after the time that the "go" decision was made at Guayamas, the spacecraft began to drift in right yaw. After allowing the spacecraft to go through several cycles of drifting in yaw attitude and then being returned by the high thrust jets, the astronaut reported that he had no 1-pound jet action in left yaw. With an astronaut aboard the spacecraft, this malfunction was considered a minor problem, especially since he still had control over the spacecraft with a number of other available control systems. It should be pointed out, however, that without an astronaut aboard the spacecraft, this problem would have been very serious in that excessive amounts of fuel would have been used; and it may have been necessary to reenter the spacecraft in some contingency recovery area because of this high fuel-usage rate.

During the pass over the control center on the second orbit, it was noticed that the telemetry channel used to indicate that the landing bag was deployed was showing a readout which, if true, indicated that the landing-bag deployment mechanism had been actuated. However, because there was no indication to the astronaut and he had not reported hearing any unusual noises or noticed any motions of the heat shield, it was felt that this signal, although a proper telemetry output, was false and probably had resulted from the failure of the sensing switch. Of course, this event caused a great deal of analysis to result and later required the most important decision of the mission to be made.

The flight continued with no further serious problems and the astronaut performed the planned 180° yaw maneuver over Africa to observe the earth and horizon while traveling in this direction and to determine his ability to control. Following this maneuver, the astronaut began to have what appeared to be trouble with the gyro reference system, that is, the attitudes as indicated by the spacecraft's instruments did not agree with the visual reference of the astronaut. However, the astronaut reported he had no trouble in maintaining the proper attitude of the spacecraft when he desired to do so by using the visual reference. Because of the problems with the automatic control system, previously mentioned, and the apparent gyro reference problem, the astronaut was forced to deviate from the flight plan to some extent, but he was able to continue all of the necessary control systems tasks and checks and to make a number of other prescribed tests which allowed both the astronaut and the ground to evaluate his performance and the performance of the spacecraft systems. As observed by the ground and the astronaut, the horizon scanners appeared to deteriorate when on the dark side of the earth; but when the spacecraft again came into daylight the reference system appeared to improve. However, analyses of the data subsequent to the flight proved that the horizon scanner system was functioning properly but the changes in spacecraft attitudes that resulted from the maneuvers performed by the astronaut caused the erroneous outputs which he noticed on the attitude instruments. It has been known that spurious attitude outputs would result if the gyro reference system were allowed to remain in effect during large deviations from the normal orbit attitude of 0° yaw, 0° roll and 34° pitch, and this was apparently the case during the 180° yaw maneuver which was conducted over Africa. This condition will be alleviated in future flights by allowing the astronaut to disconnect the horizon scanner slaving system and the programmed precession of the tyros which preserves the local horizon to be disconnected while he is maintaining attitudes other than the normal spacecraft orbit attitude.

As the go-no-go point at the end of the second and beginning of the third orbit approached, it was determined that although some spacecraft malfunctions had occurred, the astronaut continued to be in excellent condition and had complete control of the spacecraft. He was told by the Hawaiian site that the Mercury Control Center had made the decision to continue into the third orbit. The astronaut concurred, and the decision was made to complete the three orbit mission.

One other problem which caused some minor concern was the increase in inverter temperatures to values somewhat above those desired. It appeared, and the flight test results confirmed, that the cooling system for these inverters was not functioning. However, recent tests made previous to the flight had shown that the inverters could withstand these and higher operating temperatures. The results of these tests caused the flight control people to minimize this problem, and it was decided that this minor malfunction was not of sufficient magnitude to terminate the flight after the second orbit. Furthermore, a backup inverter was still available for use had one of the main inverters failed during the third orbit.

During the third orbit, the apparent problems with the gyro reference system continued and the automatic stabilization and control system (ASCS) malfunctions in the yaw axis were still evident. However, these problems were not major and both the ground and the astronaut considered that the entire situation was well under control. This was primarily because of the excellent condition of the astronaut and his ability to use visual references on both the dark and light sides of the earth, and the fact that most of the control systems were still performing perfectly. The one problem which remained outstanding and unresolved was the determination of whether the heat-shield deployment mechanism had been actuated or whether the telemetry signal was false due to a sensing switch failure. During the pass over Hawaii on the third orbit, the astronaut was asked to perform some additional checks on the landing-bag deployment system. Although the test results were negative and further indicated that the signal was false, they were not conclusive. There were still other possible malfunctions and the decision was made at the control center that the safest path to take was to leave the retropackage on following retrofire. This decision was made on the basis that the retropackage straps attached to the spacecraft and the spacecraft heat shield would maintain the heat shield in the closed position until sufficient aerodynamic force was exerted to keep the shield on the spacecraft. In addition, based on studies made in the past, it was felt that the retention of the package would not cause any serious damage to the heat shield or the spacecraft during the reentry and would burn off during the reentry heat pulse.

Also during the pass over the Hawaiian site, the astronaut went over his retrosequence checklist and prepared for the retrofire maneuver. It was agreed that the flight plan would be followed and that the retrofire maneuver would take place using the automatic control system, with the astronaut prepared to take over manually should a malfunction occur. Additional time checks were also made over Hawaii to make sure that the retrofire clock was properly set and synchronized to provide retrofire at the proper moment. The astronaut himself continued to be in excellent condition and showed complete confidence in his ability to control any situation which might develop.

The retrofire maneuver took place at precisely the right time over the California site and, as a precautionary measure, the astronaut performed manual control along with the automatic control during this maneuver. The attitudes during retrofire were held within about 3° of the nominal attitudes as a result of this procedure, but large amounts of fuel were expended. Following this maneuver, the astronaut was instructed to retain the retropackage during reentry and was notified that he would have to retract the periscope manually and initiate the return to reentry attitude and the planned roll rate because of this interruption to the normal spacecraft sequence of events.

Following the firing of the retrorockets and with subsequent radar track, the real-time computers gave a predicted landing point. The predictions were within a small distance of where the spacecraft and astronaut were finally retrieved. As far as the ground was concerned, the reentry into the earth's atmosphere was entirely normal. The ionization blackout occurred within a few seconds of the expected time and although voice communications with the astronaut were lost for approximately 4 minutes and 20 seconds, the C-band radar units continued to track throughout this period and provided some confidence that all was well throughout the high heating period. As it might be expected, voice communications received from the astronaut following the ionization blackout period resulted in a great sigh of relief within the Mercury Control Center. The astronaut continued to report that he was in excellent condition after this time, and the reentry sequence from this point on was entirely normal. A number of spacecraft control problems were experienced following peak reentry acceleration primarily because of the method of control used during this period. In addition, large amounts of fuel from both the manual and automatic systems had been used and finally resulted in fuel depletion of both systems just previous to the time that the drogue chute was deployed. The results of these flight tests have indicated that somewhat different control procedures be used during this period for the next flight. The communications with the astronaut during the latter stages of descent on both the drogue and main parachutes were excellent and allowed communications with either the astronaut or the recovery forces throughout this entire descent phase and the recovery operations which took place following the landing. The landing occurred at 2:43 p.m. e.s.t. after 4 hours, 55 minutes, and 23 seconds of flight. The recovery operations are described in paper 7.

7. RECOVERY OPERATIONS

By ROBERT F. THOMPSON, Flight Operations Division, NASA Manned Spacecraft Center; and ENOCH M. JONES, Flight Operations Division, NASA Manned Spacecraft Center

Summary

Astronaut Glenn and his spacecraft were recovered by the destroyer USS Noa in the North Atlantic planned recovery area after a flight of three orbits around the earth. A description of the events occurring in this recovery operation and a general description of the scope of recovery support required for the MA-6 flight are presented. Also, the composition and deployment of the recovery forces provided for various landing situations are outlined and the location and retrieval techniques available to these forces are discussed.

Introduction

This paper presents a general description of the total scope of recovery support provided for the MA-6 flight, briefly describes the location and retrieval techniques available in various landing situations, and describes the MA-6 recovery operation in the actual landing area. Recovery operations are defined as the support required for location and retrieval of the astronaut and spacecraft subsequent to landing. Before the MA-6 recovery is discussed specifically, two general points are noted. The support provided for all Mercury flights reflects a consideration of both normal flights and various possible abort situations; and it is the latter case, that is, supporting possible abort situations having a reasonable probability of occurring, that imposes by far the greatest support requirements on recovery forces. Consequently, while a relatively large number of recovery vehicles and personnel are required to provide the desired support capability, only a few actually become directly involved in the recovery for any given operation.

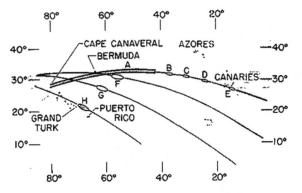

FIGURE 7-1.—MA-6 planned recovery areas.

Secondly, the recovery forces which have supported Project Mercury flight operations the airplanes, ships, helicopters, and other special vehicles, are provided by the Department of Defense, and for the most part represent operational units that devote only a relatively small part of their total workload to Mercury recovery. Recovery techniques and equipment have been developed which permit the Department of Defense to support this program with an acceptable diversion from their normal functions.

Deployment of Support Forces

In order to describe the recovery support provided for the MA-6 flight, recovery areas are considered in two broad categories: Planned recovery areas in which the probability of landing was considered sufficiently high to require the positioning of location and retrieval units assuring recovery within a specific time; and contingency recovery areas in which the probability of landing was considered sufficiently low to require only the utilization of specialized search and rescue procedures.
The planned recovery areas were all located in the North Atlantic Ocean as shown in figure 7-1, and table 7-1 is a summary of the support positioned in these areas at launch time for the MA-6 flight.

Special recovery teams utilizing helicopters, amphibious vehicles, and salvage ships were located at the launch site to provide rapid access to the spacecraft for landings resulting from possible aborts during the late countdown and the early phase of powered flight. Winds at the launch site were measured and the locus of probable landing positions for various abort times were computed to facilitate positioning of these recovery forces.

Areas A to E supported all probable landings in the event an abort was necessitated at any time during powered flight. Area A would contain landings for abort velocities up to about 24,000 feet per second, and Areas B, C, D, and E would support higher abort velocities where programmed use of the retrorockets become effective in localizing the landing area. Forces as shown in table 7-1 were positioned in these areas to provide for location and retrieval within a maximum of 3 hours in the areas of higher landing probability and 6 hours in the areas where the probability of landing was somewhat lower.

Once the spacecraft was in orbital flight, Areas F, G, and H were available for landing at the end of the first, second, or third orbits, respectively. Forces as shown in table 7-1 were available to assure location and retrieval within a maximum of 3 hours for most probable landing situations. Thus, to assure short-time recovery for all probable aborts that could occur during powered flight and for landings at the end of each of the three orbits, a total of 21 ships, 12 helicopters, and 16 search aircraft were on station in the deep-water landing areas at the time of the MA-6 launch. Backup search aircraft were available at several staging locations to assure maintaining the airborne aircraft listed in table 7-1. These forces in the planned recovery areas were all linked by communications with the recovery control center located within the Mercury Control Center at Cape Canaveral. Since it was recognized that certain low probability situations could lead to a spacecraft landing at essentially any point along the ground track over which the spacecraft flies, suitable recovery plans and support forces were provided to cover this unlikely contingency. In keeping with the low probabilities associated with remote landings, a minimum type of support was planned for contingency recovery; however, a large force is required because of the extensive areas covered in three orbits around the earth. The location of contingency recovery units for the MA-6 flight is shown in figure 7-2. A typical unit consists basically of two search aircraft specially equipped for UHF/DF homing on spacecraft beacons, point to-point and ground-to-air communications, and pararescue personnel equipped to provide on-scene assistance on both land and water. No retrieval forces were deployed in support of contingency landings; procedures were available for retrieval support for after the fact. These search and rescue units were stationed at the 16 locations shown in figure 7-2 and were all linked by communications with the recovery control center at Cape Canaveral. Throughout the MA-6 flight, the astronaut was continually provided with retrofiring times for landing in favorable contingency recovery areas. However, the contingency forces deployed had the capability of flying to any point along the orbital track if required.

TABLE- 7-1 — MA-6 Recovery Forces for the Planned Recovery Areas				
Area	Number of search aircraft	Number of helicopters	Number of ships	Maximum recovery time, hr
Launch site		3		Short
A	6		8 destroyers	3 to 6
B, C, D, E	1 each		1 destroyer each	3 to 6
F, G, H	2 each	3 each	1 carrier each & 2 destroyers each	3
Total	16	12	21	

Recovery Techniques

In order to complete the description of the recovery forces that were deployed for the MA-6 flight, it is important to have a general understanding of the techniques that were available for location and retrieval in various situations.

Location

Launch-site forces are expected to have visual contact with the spacecraft should an abort occur in their area. Launches are scheduled after daylight in the launch area and satisfactory weather is assured before launching. The launch-site recovery commander is airborne in a helicopter behind the launch pad at the time of launch, and other launch-type support forces are prebriefed and deployed where possible aborts could be effectively observed and retrieval assignments executed.

In the deep-water areas of the North Atlantic, search aircraft are airborne in the recovery areas prior to spacecraft fly-over or landing. If required, the aircraft would be directed toward a search datum established from landing-prediction information provided by the Mercury tracking-computing network and other spacecraft location systems; such as, SOPAR, the sound fixing and ranging system which utilizes an underwater detonation technique, and HF/ DF (high frequency/direction finding), the fixing of spacecraft position by land-based stations utilizing HF radio signals radiated from the spacecraft subsequent to landing. Accuracy of the datum is expected to be sufficient to bring these airborne search aircraft equipped with special electronic receivers within range of spacecraft electronic beacons operating in the UHF (ultra high frequency) range. When within UHF/DF range, the search aircraft can commence "homing" on the spacecraft electronic beacons until visual contact is established. Fluorescein sea marker and a flashing light are provided as visual location aids. Lookouts aboard the recovery ships are also on alert during the reentry and landing phase in an attempt to sight the spacecraft.

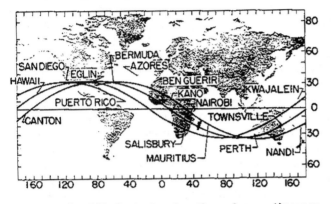

FIGURE 7-2.—MA-6 staging locations for contingency recovery.

In contingency areas, procedures utilized for MA-6 called for the search aircraft to remain on the ground in an "alert" status. In the event of a contingency landing a search area would be established by the recovery control center from information similar to that used in the planned areas for establishing the search datum. Contingency aircraft were also equipped with UHF/DF equipment compatible with spacecraft beacons and would utilize this equipment to locate the spacecraft. This "ground alert" procedure assured maximum utilization of all aircraft and still permitted reaching all possible landing sites well within the lifetime of spacecraft location aids.

Retrieval

FIGURE 7-3.—Helicopter retrieval technique.

Recovery units deployed to provide a retrieval capability generally have several techniques or modes of operation available for adapting to different possible situations. It is beyond the scope of this discussion of MA- 6 recovery operations to describe all of the various capabilities involved, and only limited comments are made in order to provide a general feeling of what retrieval support was available at the time of the MA-6 launch. All recovery ships have the basic capability of hoisting the Mercury spacecraft clear of the water and securing it on deck. Basic plans call for the astronaut to remain in the spacecraft until it is aboard ship and egress at this time; however, he could egress from the spacecraft prior to pickup if this procedure became desirable.

The helicopters which were in the launch site and in each of the end-of-orbit landing areas have three techniques available for retrieval. They are: deployment of a flotation collar and retrieval of the astronaut only, simultaneous retrieval of the astronaut and the spacecraft with transfer of the astronaut to the helicopter, and simultaneous retrieval of the astronaut and the spacecraft with the astronaut remaining in the spacecraft. Only the first case is discussed since this method of retrieval was planned as the primary technique for use by helicopters had they become involved in MA-6. Two swimmers are deployed into the water from the retrieval helicopter and they affix a flotation collar to the spacecraft as shown in figure 7-3. This collar is positioned about the spacecraft before inflation; and when the dual rings are inflated, the spacecraft is par-tially supported and, relative to the spacecraft, becomes a very stable working platform. After collar installation, the astronaut can egress either through the tower as shown in figure 7-3 or through the side hatch for transfer to the helicopter by personnel hoist. The flotation collar is also utilized in providing contingency recovery forces with an on-scene assistance capability for remote water landings. This mode of operation is depicted in figure 7-4 which shows pararescue personnel at the spacecraft following their deployment by parachute from a contingency search aircraft.

This discussion of the recovery forces that were deployed and the brief résumé of the procedures to be utilized for location and retrieval in various situations provide a background for describing the actual MA-6 location and retrieval.

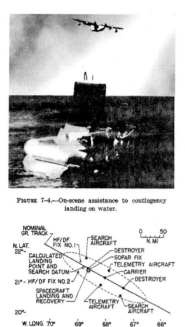

FIGURE 7-4.—On-scene assistance to contingency landing on water.

Description of MA-6 Recovery

Recovery forces in all areas were notified of mission progress by the recovery control center. Consequently, units located at the end of the third orbit knew they were to become involved, and figure 7-5 presents recovery details in the MA-6 landing area. An aircraft carrier with retrieval helicopters was located in the center of the planned landing area, one destroyer was located about 40 nautical miles downrange, and a second destroyer was located about 40 nautical miles uprange. Telemetry and search aircraft were airborne in the areas as shown. After the retrorocket maneuver and about 15 minutes prior to the estimated time of landing, the recovery control center notified the recovery forces that according to calculations, the landing was predicted to occur near the uprange destroyer as shown in figure 7-5. The astronaut was also provided with this information by the Mercury Control Center as soon as communications were reestablished after the spacecraft emerged from the ionization blackout. Lookouts aboard the USS Noa, the destroyer in the uprange position, sighted the main parachute of the spacecraft as it descended below a broken cloud layer at an altitude of about 5,000 feet from a range of approximately 5 nautical miles. Communications were established between the spacecraft and the destroyer, and a continuous flow of information was passed throughout the remainder of the recovery operation.

FIGURE 7-5.—MA-6 landing area.

In this case, location was very straightforward in that a retrieval ship gained visual contact during spacecraft landing. However, as a matter of interest for future operations since visual sightings are probably the exception rather than the rule, other spacecraft location information available soon after landing is also plotted in figure 7-5. The SOFAR fix was approximately 4 nautical miles from the landing point, and the first two HF/DF fixes were within approximately 25 miles of the actual spacecraft position. This landing information, along with the calculated landing position provided by the Mercury network, would have assured bringing search aircraft within UHF/ DF range. In fact, the airborne search aircraft in the MA-6 landing area obtained UHF/DF contact with the spacecraft shortly after beacon activation at main parachute opening; however, it was the Noa's day and she was on her way to retrieve.

The Noa had the spacecraft aboard 20 minutes after landing. Figure 7-6 shows the spacecraft as it is being lowered to the deck. Astronaut Glenn remained in the spacecraft during pickup and after it was positioned on the ship's deck, he egressed from the spacecraft through the side hatch. Original plans had called for egress through the top at this time; however, the astronaut was becoming uncomfortably warm and decided to leave by the easier egress path.

In making the pickup, the Noa maneuvered alongside the spacecraft and engaged a hook into the spacecraft's lifting loop. The hook is rigged on the end of a detachable pole to facilitate this engagement and the lifting line is rigged over one of the ship's regular boat davits as shown in figure 7-6. A deck winch is used for inhauling the lifting line, and when the spacecraft is properly positioned vertically, the davit is rotated inboard to position the spacecraft on deck. A brace attached to the davit is lowered over the top of the spacecraft to prevent swinging once the spacecraft is clear of the water.

Each ship in the recovery force had embarked a special medical team consisting of two doctors and one technician to provide medical care and/or initial postflight medical debriefing. For the MA-6 mission, postflight medical debriefing was the only requirement and was completed onboard the Noa in about 2 hours after pickup. The astronaut was then transferred to the aircraft carrier for further transfer to Grand Turk Island, and he arrived there approximately 5 hours after landing. Additional engineering and medical debriefings were conducted at Grand Turk.

FIGURE 7-6.—MA-6 retrieval by destroyer USS *Noa*.

8. AEROMEDICAL PREPARATION AND RESULTS OF POSTFLIGHT MEDICAL EXAMINATIONS

By HOWARD A. MINNERS, M.D., Life Systems Division, NASA Manned Spacecraft Center; WILLIAM K. DOUGLAS, M.D., Astronaut Flight Surgeon, NASA Manned Spacecraft Center; EDWARD C. KNOBLOCK, Ph. D., Walter Reed Army Institute of Research; ASHTON GRAYBIEL, M.D., USN School of Aviation Medicine, Pensacola, Fla; and WILLARD R. HAWKINS, M.D., Office of the Surgeon General, Hq. USAF, Washington, D.C.

Summary

The preflight and postflight medical evaluations have revealed no adverse effect of 4½ hours of space flight per se. In an effort to interpret such normal results, three alternatives come to mind:

(1) As measured by available techniques of examination, space flight has, indeed, no ill effect.

(2) The effects of space flight may be so evanescent that they were resolved before the pilot could be examined after the flight.

(3) The MA-6 space flight was of insufficient duration to produce detectable effects or such effects have not yet become evident.

Further study in future manned space flights should help to determine which of these, or other, possible interpretations is correct.

Introduction

Comprehensive medical evaluations of Astronaut John H. Glenn, Jr., were performed prior to his orbital space flight and as soon after his flight as recovery practices permitted. Primarily, these examinations were accomplished to determine the pilot's state of health and his medical fitness for flight. In addition, such clinical evaluations serve as baseline medical data which may be correlated with in-flight physiological information. Aeromedical data sources utilized to establish Astronaut Glenn's state of health are:

(1) Prior medical examination commencing with astronaut selection in 1959.

(2) Detailed preflight clinical evaluations conducted prior to the canceled and successful missions.

(3) Immediate preflight examination conducted on launch morning.

(4) Postflight medical examinations aboard the recovery ships and at the Grand Turk Island medical facility.

Preflight Examinations

The pilot's general preflight activities, commencing with preparation for a planned early January 1962 launch, are summarized in table 8-1. Throughout this period, his physical and mental health remained excellent.

Lift-off marks the simultaneous culmination of a number of different countdowns used for such a complex mission. The aeromedical countdown represents an effort not only to afford the pilot sufficient time for sleep and immediate preparation for the mission, but also seeks to insert him into the spacecraft at the time required by the other countdowns. By careful planning of aeromedical countdown events, the pilot can be maintained in optimum condition and embarks on the flight with a minimum of fatigue and in the best physical condition. Significant MA-6 aeromedical countdown events are listed in table 8-11. A total of 7 hours and 27 minutes elapsed between awakening the pilot and lift-off; he did not get to sleep again until 23 hours and 10 minutes after being awakened at 2:20 a.m. e.s.t. on the morning of the flight. Therefore, on flight day the pilot spent more than 5 hours on the ground for every hour in space flight.

TABLE 8-1.— MA-6 Pilot's Preflight Activities	
Date	Activity *
December 1961 13, 15, and 16	Simulated orbital mission
January 1962 15	Flight acceptance composite test, launch pad
16 and 18	Simulated orbital mission
17	Launch simulation, launch pad; network simulation
19	Launch simulation
20 and 23	Simulated flight, launch pad
22	Full mission simulation; specialists medical examinations, hospital
27	Preflight physical examination, countdown, launch pad; canceled flight
February 1962 7 and 8	Simulated launch aborts
9 and 13	Mercury orbital simulation
12	Specialists medical examinations, hospital
15	Insertions
17	Launch simulation
20	Actual MA-6 flight
* In Cape Canaveral procedures trainer, unless otherwise stated.	

Figure 8-1.—Balance test.

TABLE 8-II.— Significant Events Prior to Launch		
Date	e.s.t.	Event
February 16,1962	a.m.	Began low residue diet
February 19, 1962	9:30 p. m.	Retired *
February 20, 1962	2:20 a.m.	Awakened and showered
	2:50 a.m.	Breakfast
	3:05 a.m.	Physical examination
	4:28 a.m.	Suiting started
	5:05 a.m.	Entered transfer van
	5:20 a.m.	Arrived at launch pad and remained in transfer van
	5:58 a.m.	Ascended gantry
	6:06 a.m.	Insertion into spacecraft
	6:25 a.m.	Countdown resumed
	9:47 a.m.	Launch
* Obtained 4 hours and 50 minutes of dozing, light sleep. No medication administered.		

Detailed medical examinations were conducted prior to the canceled flight in January 1962 and before the flight in February. Aspects of the examination which were not time critical were completed several days before the launch and included the following: specialists' evaluations in neurology, ophthalmology, aviation medicine, psychiatry, and radiology; a standard 12-lead electrocardiogram, an audiogram, and an electroencephalogram; and biochemical studies of blood and urine. All of these evaluations showed normal results and revealed no change from the numerous preceding examinations. In addition, special labyrinthine studies were performed in which the pilot was timed and scored on his ability to maintain his balance while walking along successively more narrow rails as depicted in figure 8-1. Astronaut Glenn's routine scores on these rails are considerably higher than those which have been obtained from a group of flight personnel. Also, his auditory canals were irrigated for 45 seconds with carefully temperature regulated water, and the warmest temperature at which fine nystagmus began was recorded as the threshold temperature. The pilot was again examined on launch morning by specialists in aviation medicine and internal medicine. This examination, begun at 3:05 a.m. e.s.t. on February 20, 1962 (the day of the flight), revealed a calm, healthy, and alert adult male. Vital signs were as follows: pulse, 68 beats per minute and regular; blood pressure 118/80 mm Hg (left, arm sitting) ; respiration 14 breaths per minute; oral temperature, 98.20 F; and nude weight with the bladder empty 171 pounds 7 ounces. Eyes, ears, nose, and throat were normal and unchanged from previous examination. Lungs were clear throughout and diaphragmatic excursion was full. Examination of the heart revealed that the aortic second sound was equal to the pulmonic second sound when the examinee was in a sitting position, and the pulmonic second sound was greater than the aortic second sound when the examinee was supine. The cardiac sounds were of good quality, were not split, and there were no murmurs. The abdomen was well relaxed; there was no tenderness and no masses. The liver edge was barely palpable; and there was no costovertebral angle nor bladder tenderness. Skin was clear and a cursory neurologic examination was normal. Some of these findings, along with extremity measurements, are listed in table 8-III. The authors have taken the liberty of summarizing the clinical findings of Dr. Myers, Neurologist, Dr. Clark, Ophthalmologist, Dr. Ruff, Psychiatrist, and Drs. McIver and Mullin, recovery forces physicians.

TABLE 8-III.— Clinical Evaluation
(All times are Eastern Standard)

	Preflight (launch morning)	Postflight
General status	Eager for flight	Alert, but not talkative; sweating profusely; appeared fatigued; not hungry.
Weight, lb	171 7/16 at 3:15 a.m.	166 2/16 at 6:50 p.m. (5 5/16 lb loss)*.
Temperature, °F	98.2 (oral)	99.2 (rectal at 4:00 p.m.); 98.0 (oral at 12:00 p.m.).
Respiration, breaths/min	14	14
Pulse, beats/min	68	76 on shipboard, 72 at Grand Turk.
Blood pressure, (left arm), mm Hg.	118/80 (sitting)	105/60 (standing); 120/60 (supine) at 3:45 p.m.; 128/78 (sitting) at 9:30 p.m.
Heart and lungs	Normal	Normal — no change.
Skin	No erythema or abrasions	Erythema of biosensor sites; superficial abrasions second and third fingers of right hand.
Extremity measurements:	Left Right	Left Right
Wrist, in	6 7/8 7	6 3/4 7
Calf (maximum) (in)	16 7/8 16 1/2	16 5/8 16 1/8
Ankle (minimum) (in)	9 3/8 9 1/8	9 9 1/4

* Not true inflight weight loss since the scales were neither the same nor compared and postflight weight was 4 hours, 8 minutes after landing.

TABLE 8-IV.— *Xylose Absorption Study*

Subject	Xylose recovered, percent, after —														
	1 hr			2 hr			3 hr			4 hr			5 hr		
	For test condition —														
	a	b	c	a	b	c	a	b	c	a	b	c	a	b	c
1	13.7	----	17.6	26.7	24.3	30.0	36.2	35.2	36.8	----	----	41.0	41.6	43.3	45.4
2	----	----	12.2	6.7	11.3	22.0	14.3	20.0	30.3	17.0	24.8	36.7	24.4	28.6	39.0
3	----	----	18.1	15.4	20.1	31.1	24.7	29.1	38.5	31.7	35.3	42.4	32.9	38.8	45.5
4	----	----	15.3	24.2	24.2	23.1	32.3	30.9	26.7	36.2	34.7	30.9	38.8	36.8	-----
5	-----		9.6	18.1	20.5	19.2	24.8	29.1	26.3	31.0	34.5	30.9	35.2	38.7	33.3
JHG-control *	----	7.1	....	----	21.9	----	----	29.7	----	---	38.2	----	----	42.1	-----
JHG-flight **	....	....	....	....	....	....	....	....	....	....	....	34.9	----	----	-----

The test conditions were:
a- Subject fasting, xylose, 5 gm, oral dose in 250 ml water. No additional fluid intake.
b- Subject fasting, xylose, 5 gm, oral tablet. Water as desired.
c- Subject, non-fasting, xylose, 5 gm, oral tablet. Test meal of applesauce and beef at 30 minutes after xylose. Water as desired.
* John H. Glenn, Jr.— Control followed condition b.
** John H. Glenn, Jr.— Conditions approximated condition c. Xylose taken T+23 minutes with voiding at approximately T+263 minutes. Applesauce only in flight.

Postflight Examinations

Postflight medical evaluation began when Astronaut Glenn emerged from the spacecraft on board the destroyer Noa 39 minutes after landing. The pilot was described as appearing hot, sweating profusely, and fatigued. He was lucid, although not talkative, and had no medical complaints other than being hot; there was no other subjective evidence of dehydration. After removal of his pressure suit and a shower, the pilot began with the shipboard medical debriefing. A brief medical history of the space flight revealed that in spite of voluntary, rather violent head maneuvers by the pilot in flight, he specifically noted no gastrointestinal, vestibular, nor disorientation symptoms while weightless. Likewise, he experienced no adverse effects from isolation or confinement. Specifically, there was no sensory deprivation. His flight plan, in addition to the requirement to control the spacecraft on the fly-by-wire system, kept Astronaut Glenn very active and busy during the flight, and there was no so-called "break-off" phenomenon. As evidenced by the numerous in-flight reports, by task performance, and by the onboard film, the pilot's mental and psychomotor responses were consistently appropriate. Psychiatrically, both before and after, and during the flight, he exhibited entirely normal behavior. He did describe a mild sensation of "stomach awareness," which in no way approximated nausea or vomiting. This sensation, which cleared spontaneously in 1½ hours, began after the spacecraft was on the water and during the 20-minute period before recovery. At the time of landing, the ambient air temperature was 76° F with 60 to 65 percent relative humidity; suit inlet temperature was 85° F; cabin air temperature was 103° F. The pilot ingested the equivalent of only 94 cubic centimeters of water (applesauce puree) for the rather long period of almost 13 hours from breakfast at 2:50 a.m. e.s.t. to shipboard at 3.45 p.m. e.s.t. During the flight, he also ate one 5.0 gram sugar tablet (xylose). Gastrointestinal function while weightless, as measured by xylose absorption, was normal. (See table 8-IV) — The fluid intake and output is shown in table 8-V.

TABLE 8-V.— Fluid Intake and Output			
	Urine e.s.t. Output ₁	Fluid Intake	e.s.t.
Countdown	0 cc	0 cc	
Inflight	₂ 800cc 2:00 p.m.	₃ 94cc	11:48 a.m.
Postflight, ship	0 cc	265cc iced tea. 240cc water 125cc coffee	3:45 p.m. 6:30 p.m. 6:50 p.m.
Total	800 cc	724 cc	

₁ See also table 8-VI.
₂ Specific gravity, 1.016.
₃ 119.5 grams of applesauce puree (78.7 percent water)

The immediate postflight medical examination onboard the destroyer recorded the following vital signs: rectal temperature, 99.2° F, blood pressure, 120/60 mm Hg supine, pulse 76 beats/min and regular. There were two small superficial skin abrasions of the knuckles of the second and third fingers of the right hand without deformation or fracture. These were received when the plunger of the explosive actuator for the egress hatch recoiled against the pilot's gloved hand. The skin also revealed an area of moderate reddening and a pressure point at the site where the blood pressure microphone had been attached. Furthermore, there was a mild reaction to the moleskin adhesive plaster which attached the four ECG electrodes. Head, eye, ear, nose, and throat examinations were normal. The heart rhythm, size,

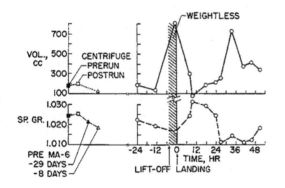

FIGURE 8-2.—Urine summary.

TABLE 8-VI.— Urine Summary																	
	Centrifuge		Preflight			MA-6 flight		Postflight									
	Prerun	Postrun +2 hr	-29 days	-8 days	-2 days	03:30 a.m. c.s.t. flight day	03:30 a.m. to 02:10p. m e.s.t. inflight	+8 hr	+10 hr	+18 hr	+24 hr	+27 hr	+34 hr	+41 hr	+46 hr	+51 hr
Volume, cc	180	195	-----	120	185	135	900	295	76	192	210	250	720	365	405	335
Specific gravity	1.024	1.025	1.021	1.018	1.022	1. 019	1. 016	1.024	1.031	1.029	1.024	1.011	1.014	1.011	1.012	1.018
Albumin	Neg.	Neg.	Neg.	Neg.	Neg.	Neg.	Neg.	Neg.	Neg.	Neg.	Neg.	Neg.	Neg.	Neg.	Neg.	Neg.
Glucose	Neg.	Neg.	Neg.	Neg.	Neg.	Neg.	Neg.	Neg.	Neg.	Neg.	Neg.	Neg.	Neg.	Neg.	Neg.	Neg.
Ketones	Neg.	Neg.	Neg.	Neg.	Neg.	Neg.	Neg.	Neg.	Neg.	Neg.	Neg.	Neg.	Neg.	Neg.	Neg.	Neg.
pH	6.3	6.2	6.0	6.1	6.0	6.2	6.0	6.0	6.0	----	----	----	----	----	----	------
Na, mEq/L	123	155	-----	-----	------	225	157	103	89	96	88	61	73	49	125	145
K, mEq/L	95	97	-----	-----	------	64	27	59	66	38	35	17	15	23	41	41
Cl, mEq/L	-----	-----	-----	-----	------	223	152	100	30	20	77	45	67	39	140	141
Ca, mEq/L	-----	-----	-----	-----	------	11.2	6.9	3.5	4.9	15.8	24	1	3	5.9	7.7	10.6
Osmolarity (milliosmoles)	-----	-----	-----	-----	------	1010	613	599	920	1111	1000	432	460	511	590	720
Microscopic examination	No formed elements.				0cc. Squamous Epithelial cell.				No formed elements.							

TABLE 8-VII.— *Peripheral Blood*							
	Preflight					Postflight	
	Mar. 1959	Aug 1966	Aug. 1961	-29 days *	-8 days *	+8hr *	+46 hr *
Hematocrit, percent	45	45	42	-------	39.5	46	42
Hemoglobin (Cyanmethemoglobin method), grams/100 ml	15.7	15.3	13.6	14.5	14.1	16.1	14.7
Red blood cells X 10⁶ / mm³	-------	-------	-------	4.75	4.96	4.82	5.03
White blood cells/mm³	5,000	5,000	6,310	5,100	4,650	8,200	5,450
Differential white-blood count: Lymphocytes, percent	40	41	45	37	47	36	33
Neutrophiles, percent	58	49	42	57	47	58	57
Monocytes, percent	1	6	7	3	3	3	3
Eosinophiles, percent	1	4	5	1	2	2	5
Basophiles, percent	1	0	1	2	1	1	2
*Determinations by same technician.							

and sounds were normal, and the lungs were clear, without physical evidence of local pulmonary collapse. Abdominal examination was negative, and the lower extremities showed no swelling nor evidence of venous thrombosis. A general neurologic examination was entirely normal. Urine and blood samples were obtained for later analysis, (See tables 8-VI and 8-VII and fig. 8-2.) The in-flight urine collection device contained 800 cubic centimeters of clear, straw-colored urine with a specific gravity of 1.016, pH 6.0, and was negative microscopically and for blood, protein, glucose and acetone. This volume of urine was passed just prior to retrosequence; bladder sensation and function while weightless was normal and unchanged from that of the customary 1g, ground environment.

At 5:45 p.m. e.s.t. the pilot was transferred to the aircraft carrier U.S.S. Randolph where posterior-anterior, and lateral chest X-rays, standard 12-lead electrocardiogram, and body weight were obtained. Body weight could not be obtained sooner due to the rolling of the destroyer.

Later, Astronaut Glenn flew to Grand Turk Island where a general physical examination was begun at 9:30 p.m. e.s.t. approximately 6¾ hours after spacecraft landing. The vital signs at that time were: blood pressure, 128/78 mm Hg (left arm, sitting) ; pulse, 72 beats/ min; respiration 14 breaths/min; and oral temperature, 98° F at 12:00 p.m. e.s.t. Except for the previously described superficial skin abrasions, the entire examination was normal. During the subsequent 48 hours, a comprehensive examination was conducted by the same medical specialists who examined the pilot prior to flight. The psychiatrists examination was entirely normal, a neurologic examination, including an electroencephalogram, was unchanged except for slightly increased deep tendon reflexes, and the ophthalmologist found normal and unchanged ocular function, without slit lamp or funduscopic evidence of cosmic ray damage. The internist noted no change in the lung fields nor in the quality and character of the cardiac sounds. The preflight and postflight chest X-rays and electrocardiograms were compared; neither abnormality nor change was observed. The identical, special labyrinthine tests were performed in an effort to demonstrate any effect of space flight upon the pilot's sense of balance. These postflight labyrinthine tests, and both the general and specialists' examinations, revealed no significant changes from the pilot's preflight condition.

There were, however, a few measured differences between the preflight and postflight medical findings, and these are summarized in table 8-III. The pilot lost only slightly more weight than he lost during a Mercury-Atlas three-orbit simulation on the centrifuge. Such weight loss, coupled with a diminished urine volume and increased specific gravity after the flight (see table 8-VI and fig. 8-2), hemoconcentration (see table 8-VII), and the recovery physicians clinical impression led to the diagnosis of mild dehydration. The battery of biochemical studies (see tables 8-VIII and 8-IX) further supports this impression; however, no abnormality specifically attributable to space flight, as opposed to atmospheric flight, is evident. Astronaut Glenn further reported no subjective symptoms of dehydration other than being hot. His mild dehydration was due primarily to the overheating experienced just prior to landing and while on the water awaiting pickup. Also, he had a minimal fluid intake for almost 13 hours; his intake and output is summarized in table 8-V. His mild postflight gastrointestinal "awareness" is also attributed to the increased environmental temperature after the flight and to mild dehydration. In addition, the bobbing motion of the spacecraft on the sea is undoubtedly a major contributory factor.

TABLE 8-VIII.— Blood Summary *	Centrifuge			MA-6 flight				
	Prerun	Postrun		Preflight		Postflight		
		+2 hr	+6hr	-29 days	-8 days	+1 hr	+8 hr	+46 hr
Glucose (whole blood) mgm/100 ml	97	112	121	95	109	-	96	99
Sodium (serum), mEq/L	143	140	154	155	160	146	144	143
Potassium (serum), mEq/L	4.8	4.8	5.6	5.4	4.6	3.9	4.4	4.4
Calcium (serum), mEq/L	5.2	6.0	5.2	4.9	4.3	4.3	4.2	4.4
Chloride (serum), mEq/L	80	83	83	95	98	104	104	104
Protein (total serum), g/100 ml	7.9	7.7	8.0	6.9	6.6	6.9	6.6	6.7
Albumin (serum), g/100 ml	4.3	4.1	4.7	4.1	3.8	3.8	3.8	3.8
Albumin/Globulin ratio (serum)	1.2	1.1	1.2	1.4	1.4	1.2	1.4	1.3
Urea Nitrogen (serum) mg/100 ml	15.4	16.0	14.3	14.1	15.5	10.5	10.5	11.6
Epinephrine, plasma μg/L	<0.1	<0.1	<0.1	<0.1	<0.1	<0.1	<0.1	<0.1
Norepinephrine, plasma μg/L	<0.1	<0.1	<0.1	5.0	18.0	-	6.0	3.8
* Operational priorities precluded making a biochemical requirement for fasting specimens.								

TABLE 8-IX.— Plasma Enzymes Summary

	Normal values	Centrifuge				MR-3 backup	MR-4 backup	MA-6 flight			
		Prerun	Postrun +6 hr	Prerun	Postrun +2 hr			Pre-flight	Postflight		
									+1 hr	+8 hr	+46 hr
Transaminases:											
SCOT	0 to 35	19	27	68	62	33	48	18	33	27	23
SGPT	0 to 20	6	10	50	50	10	4	----	----	----	----
Esterase:											
Acetylcholine	130 to 260	165	185	----	----	250	220	240	305	245	275
Peptidase:	100 to 310	220	250	350	350	390	370	300	250	290	255
Leucylamino	--------	----	----	250	150	----	----	----	----	----	----
Leucylamino, heat stable											
Aldolase	50 to 150	25	----	50	40	41	13	112	90	209	120
Aldolase, heat treated	0	----	----	----	----	----	----	22	40	35	25
Isomerase:	10 to 20	10	9	30	51	13	23	0	24	22	9
Phosphohexase											
Dehydrogenases:											
Lactic	150 to 250	190	125	560	860	250	235	265	220	390	250
Lactic, heat stable	--------	----	----	375	595	----	----	95	60	95	85
Percent residual	14 to 15	----	----	----	----	----	----	39.6	19.7	38.7	30.9
Malic	150 to 250	250	----	530	905	225	280	----	----	----	----
Malic, heat stable	--------	----	----	330	420	----	----	----	----	----	----
Succinic	0	0	0	0	0	0	0	----	----	----	----
Inoshie	0	0	0	22	15	5	3	----	----	----	----
Alpha-ketoglutaric	0	0	0	----	----	----	----	----	----	----	----
Beta-glutamic	0	0	0	----	----	10	----	----	----	----	----
Phosphatase:											
Alkaline	10 to 20	----	----	11	0	8	5	----	----	----	----
Lactic Acid	--------	----	----	----	----	----	----	36	185	88	55
Cholesterol	-------	----	----	----	----	----	----	197	----	----	185
Cholesterol esters, percent	--------	----	----	----	----	----	----	70	----	----	71
DPNH Oxidation (non-specific enzyme activity)	0							26	65	92	46

Acknowledgements.-Special acknowledgment is paid to the following for their assistance in the medical studies: Paul W. Myers, M.D., and Charles C. Watts, Jr., M.D., Lackland Air Force Hospital, San Antonio, Texas; W. Bruce Clark, M.D., USAF School of Aerospace Medicine, San Antonio, Texas; George Ruff, M.D., University of Pennsylvania; Walter Frojola, Ph. D., Department of Biochemistry, Ohio State University; Kristen B. Eik-Nes, M.D., University of Utah; Hans Weil-Malherbe, M.D., St. Elizabeth's Hospital, Washington, D.C.; Leonard Laster, M.D., National Institutes of Health; and S/Sgt. Carlton L. Stewart, Lackland Air Force Hospital, San Antonio, Texas.

Bibliography

DOUGLAS, WILLIAM K.. JACKSON, CARMAULT R.. Jr., et al.: Results of the MR-4 Preflight and Postflight Medical Examination Conducted on Astronaut Virgil I. Grissom. Results of the Second U.S. Manned Suborbital Space Flight July 21, 3.961. NASA Manned Spacecraft Center.

JACKSON, CARMAULT B., Jr.. DOUGLAS, WILLIAM K., et al.: Results of Preflight and Postflight Medical Examination. Proc. Conf. on Results of the First U.S. Manned Suborbital Space Flight. NASA, Nat. Inst. Health, and Nat. Acad. Sci. June 6, 1961, pp. 3146.

Glucose:
NELSON, M.: Photometric Adaptation of Somogyi Method for Determination of Glucose. Jour. Biol. Chem., vol. 153,1944, pp. 375-380.

Total protein, albumin:
COHN, C., and WOLFSON, W. G..: Studies in Serum Proteins. I   The Chemical Estimation of Albumin and of the Globulin Fractions in Serum. Jour. Lab. Clin. Med., Vol. 32, 1941. pp. 1203-1207.

GORNALL, A. G., BARDAWILL, C. J., and DAVID, M. M.: Determination of Serum Proteins by Means of the Biuret Reaction. Jour.Biol. Chem., Vol. 177,1949, pp. 751-766.

Urea nitrogen:
GENTZKOW, C. J., and MASEN, J. M.: An Accurate method for the Determination of Blood Urea Nitrogen by Direct Nesslerization. Jour.Biol.Cbem.,vol.143,1942, pp. 531-544.

Calcium:
DIERL, H., and ELLINGBOE, J. L.: Indicator for Titration of Calcium in Presence of Magnesium 'With Disodium Dihydrogen Ethylene Diaminetetraacetate. Anal. Chem., Vol. 28, 1956, pp. 882-884,

Chloride:
SCHALES, O., and SCHALES, S. S.: A Simple and Accurate Method for the Determination of Chloride in Biological Fluids. Jour. Biol. Chem., VOL 140, 1941. pp. 879-884.

Epinephrine and norepinephrine:
WEIL-MALHERBE, H., and BONE, A. D.: The Adrenergic Amines, of Human Blood. Lancet, vol. 264, 1933, pp. 974- 977.

GRAY, L, YOUNG, J. G., KEEGAN, J. F., MEHAMAN, B., and SOUTHERLAND, E. W.: Adrenaline and Norepinephrine Concentration in Plasma of Humans and Rats. Clin. Chem., vol. 3, 1957, pp. 239-248.

Sodium potassium by flame photometry:
BERKMAN, S., HENRY, R. J., GOLUB, O. J., and SEAGALOVE, M.: Tungstic Acid Precipitation of Blood Proteins. Jour. Biol. Chem., Val. 206, 1954, pp. 937-943.

Vanyl mandelic acid:
SUNDERMAN, F. W., Jr., et al.: A Method for the Determination of 3-Methoxy-4-Hydroxymandelic Acid ("Vanilmandelic Acid") for the Diagnosis of Pheochromocytoma. Am. Jour. Clin. Pathol., Vol. 34, 1960, pp. 293 - 312.

Heat-stable lactic dehydrogenase:
STRANDJORD, PAUL E., and CLAYSON, KATHLEEN C.: The Diagnosis of Acute Myocardial Infarction on the Basis of Heat-Stable Lactic Dehydrogenase. Jour. Lab. Olin. Med., vol 58, 1961, pp. 962-966.

Astronaut John H. Glenn Jr.
Pilot of Mercury - Atlas 6

Climbing aboard Friendship 7 before launch (below).

The Launch of Mercury-Atlas 6 February 20th 1962 (below)

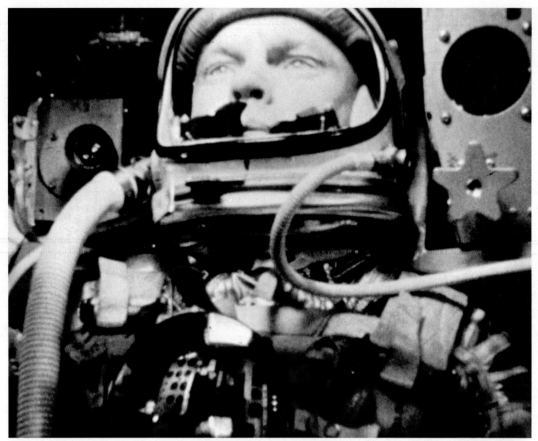

A camera onboard the "Friendship 7" Mercury spacecraft photographs John Glenn during the flight.

A tube of apple-sauce floats in front of Glenn (above), Glenn checks his mobility inside the helmet (below)

MA-6 Pre-flight activity — John Glenn and technicians inspect a decal ready for application to the side of his Mercury - Atlas 6 spacecraft prior to launch on February 20, 1962. The decal reads "Friendship 7".

The interior of the Mercury spacecraft clearly illustrating the cramped nature of the vehicle. (below and right)

The legendary Mercury Seven on desert survival training at Stead Air Force Base in Nevada. Attired in bits of parachute are (l to r) L. Gordon Cooper Jr., M. Scott Carpenter, John H. Glenn Jr., Alan B. Shepard Jr,. Virgil I. (Gus) Grissom, Walter M. Schirra Jr,. and Donald K. Slayton

The Launch console from Kennedy Space Center that was used for MA-6. Now in retirement at the Michigan Space & Science Center in Jackson Michigan. (below)

Unlike most later flights, "Friendship 7" was not equipped with a highly sophisticated camera for use by the pilot. The film used was 35mm ECN shot in a modified Ansco Autoset Camera. Glenn managed to shoot a sequence of shots of the Earth below such as this one of the Western Sahara desert (left) and this one of the Atlantic Ocean (below).

"The most impressive thing that you could see. They were brilliantly colored hues and the colors stretch out from the sun to the horizon. It is a very bright orange color down to the surface, it pales out into sort of a blue, a darker blue and then off into black, sort of like a spectrograph, same principle as the rainbow." *John H. Glenn Jr. observes sunset from orbit (below).*

An array of various delicacies concocted for project Mercury. (above) Most of the food was supplied in bags or tubes so that positive pressure could be applied by squeezing them like toothpaste.

The crew of the USS Noa use a steel cable attached to a lifeboat davit to secure Friendship 7. (above and left) The water is dyed green by a yellow floating marker device used to assist in aerial location of the spacecraft.

A cut-away view of the Mercury capsule heatshield showing not only the honey-comb of impact absorbing aluminum but also a portion of the delicate systems. (above)

Friendship 7 after being hoisted aboard the USS Noa is inspected by technicians (left).

Astronaut John H. Glenn Jr, is pulled aboard a Navy helicopter on his way from the Noa to the carrier USS Randolph. He is seen holding a bag of film which he used in orbit. (right)

During ceremonies at Launch complex 14 Glenn presents President John F. Kennedy a VIP Launch Crew Hard Hat (below).

Shortly after disembarking from his spacecraft Glenn is handed his ditty bag which contains his flight gear and the film from his voyage. It is the fourth sunset he has seen in the same day behind him. (above)

Astronaut John Glenn rides down Broadway as millions of well wishers stage a record hero welcome, for the first U.S. man to orbit the earth (above).

Two pioneers destined for the history books. During tropic survival training at Albrook Air Force Base, Canal Zone, John H. Glenn Jr and Neil Armstrong relax in their jungle abode for two nights - the "CHOCO HILTON" June 1963 (right).

9. PHYSIOLOGICAL RESPONSES OF THE ASTRONAUT

By C. PATRICK LAUGHLIN, M.D., Life Systems Division, NASA Manned Spacecraft Center; ERNEST P. MCCUTCHEON, M.D., Life Systems Division, NASA Manned Spacecraft Center; RITA M. RAPP, Life Systems Division, NASA Manned Spacecraft Center; DAVID P. MORRIS, Jr., M.D., Life Systems Division, NASA Manned Spacecraft Center; and WILLIAM A. AUGERSON, M.D., U.S. Army, Ft. Campbell, Ky.

Summary

The MA-6 mission provided a period of extended weightlessness during which the astronaut's physiological responses apparently stabilized. The values attained were within ranges compatible with normal function. No subjective abnormalities were reported by the pilot.

Introduction

The orbital space flight of Astronaut Glenn has provided a significant addition to the body of information reflecting human responses to this new situation. The life-science objectives of the flight included the study of the effects of weightlessness, launch and reentry accelerations, and weightless transition periods. The much shorter Redstone ballistic flights permitted little time in the weightless phase for physiological adjustment mechanisms to stabilize. The MA-6 mission provided a period of weightlessness of sufficient duration so that the pilot's physiological responses attained a relatively steady state. In addition to the biosensor data, the pilot's subjective evaluation of general body sensations provided a very important source of information. His comments regarding body function, such as, spatial orientation, eating, urination, and task performance, were regarded as most significant. Additional information on the operation of the spacecraft environmental control system and the bio-instrumentation system was also obtained. The approach to these studies remains essentially as outlined in the MR-3 and MR-4 postflight reports (refs. 1 and 2). Physiological investigations must comply with the overall operational requirements of the mission. There are many unexplored study areas and future flights will provide additional information.

The Space Flight Environment

The astronaut's activities during the time immediately prior to the countdown are noted in paper 8. A 38-minute hold was called after the transfer van arrived at the launch pad and the astronaut remained in the van until 5:58 a.m. e.s.t. when he ascended the gantry. Insertion into the spacecraft occurred at 6:06 a.m. e.s.t., and continuous physiological monitoring began at this time. He wore the Mercury full pressure suit and was positioned in his contour couch in the semi-supine position, with head and back raised 12° from the horizontal and hips and knees flexed at approximately 90° angles. A shoulder and lap harness secured him in the couch. During weightless flight the spacecraft was oriented so that he was in a sitting position relative to the earth's surface. At reentry he was exposed to the force of inertia in the supine position.

After completion of suit purging he was maintained on 100-percent oxygen at 14.7 psia until suit and cabin pressure declined during launch. Cabin and suit pressure regulation proceeded normally, and levels were stable at approximately 5.7 and 5.8 psia, respectively, until snorkel valve opening during the reentry sequence, after which ambient air was introduced into the system.

The total time in the spacecraft during the prelaunch period was 3 hours and 41 minutes. During this period the astronaut performed numerous spacecraft checks and prevented fatigue with frequent deep-breathing and muscle tensing exercises. Lift-off occurred at 9:47 a.m. e.s.t. and the flight proceeded as planned. Accelerations during powered flight were from 1g to 6.7g in 2 minutes and 10 seconds (booster engine cutoff, BECO), and from 1.4g to 7.7g in the following 2 minutes and 52 seconds (sustainer-engine cutoff, SECO). Spacecraft separation from the launch vehicle and the beginning of weightlessness occurred at T+5 minutes and 2 seconds. The three-orbit mission resulted in a total weightless period of 4 hours and 38 minutes.

Reentry began with 0.05g at T+4 hours and 43 minutes. Maximum reentry acceleration forces of 7.7g occurred at T+4 hours and 47 minutes with gradual buildup from 1g and gradual return to 1g over a period of 3 minutes and 30 seconds. With main parachute deployment at T+4 hours and 50 minutes, there was a brief 3.7g spike. The spacecraft landed on the water at T+4 hours and 55 minutes, 2.43 p.m. e.s.t.

Monitoring and Data Sources

Data reflecting physiological responses to flight were obtained by evaluating the biosensor real-time recordings from range stations and from the continuous onboard recording. In addition, various in-flight tests and the pilot-observer camera film were utilized for further objective analysis. The reports of values from the range medical monitors provided vital continuous coverage enabling accurate appraisal of the astronaut's status during the flight. Subjective evaluation included pilot reports from onboard voice recordings and the postflight debriefing. The countdown period provided baseline preflight information. Useful comparative measurements were available from the Mercury Atlas three-orbit centrifuge simulation, pad simulated launch, simulated flights, and the January 27, 1962, launch attempt.

Bio-instrumentation

In addition to the type of bio-instrumentation used in the manned Mercury-Redstone flights (two ECG leads, respiratory rate sensor, and body-temperature sensor), a blood pressure measuring system as described in paper 3 was utilized in flight. Preflight and postflight calibrations of the blood pressure system showed no significant change.

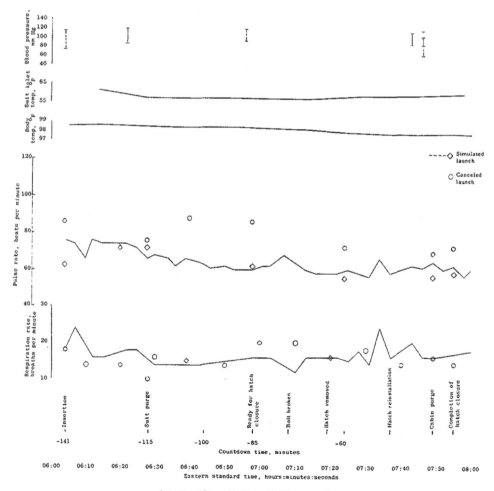

(a) Countdown 06:00 to 08:00 a.m. e.st.

FIGURE 9-1.—Preflight: Respiration rate, pulse rate, body temperature, suit-inlet temperature, and blood pressure for MA-6 countdown with values at selected events from the simulated launch of January 19, 1962, and the launch attempt of January 27, 1962.

The total biosensor monitoring time, from astronaut insertion until just prior to landing, was 8 hours and 33 minutes. The biosensor readout quality was excellent throughout the countdown and flight with the exception of the respiratory trace. As in prior flights, variation with head position and air density combined to reduce the quality of the respiration trace. There were brief periods of noise on the ECG channels during countdown and flight, usually occurring during vigorous pilot activity.

Preflight

During approximately 45 minutes in the transfer van, the astronaut's pulse ranged from 58 to 82 beats/minute with a mean of 72 and the blood pressure was 122/77 mm Hg. Figure 9-1 depicts the pulse rate, respiration rate, body temperature, suit-inlet temperature, and blood pressure values recorded during the MA-6 countdown. Values at selected events for the same physiological functions obtained from the simulated launch of January 19, 1962, and the launch attempt of January 27, 1962, are also shown. Pulse and respiration rates were determined by counting the rates for 30 seconds every 3 minutes until 10 minutes prior to liftoff when 30-second duration counts were made each minute.

The pulse rates during the launch attempt of January 27, 1962, varied from 60 to 88 beats/minute with a mean of 70 beats/minute. These rates were essentially the same as those observed during the MA-6 countdown, as shown in figure 9-1. Respiration rates were similar, varying from 12 to 20 breaths/minute. Blood pressure values from the simulated launch also approximated those observed during the MA-6 countdown. A pulse rate of 110 beats/minute and a blood pressure 139/88 mm Hg was observed at lift-off.

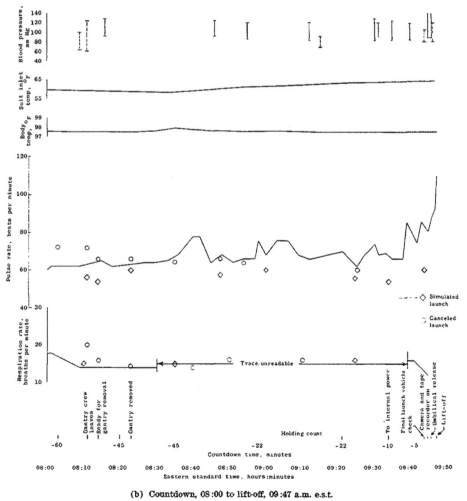

(b) Countdown, 08:00 to lift-off, 09:47 a.m. e.s.t.
FIGURE 9–1.—Concluded.

The low suit-inlet temperature maintained during countdown resulted in the pilot feeling cold and was accompanied by a fall in body temperature from 98.6° F at insertion to 97.6° F at lift-off. An examination of the electrocardiographic waveform obtained during the MA-6 countdown revealed a number of variations in the pacemaker activity which had been observed previously. These included sinus pauses, sinus bradycardia, premature atrial and nodal beats, and premature ventricular beats. On several instances some of these reported findings occurred with deep respiration. Similar variations were also recorded from the simulated launch of January 19 and from the launch attempt of January 27. In addition, a brief (16 beats) run of atrial rhythm with a rate of 100 beats/minute occurred during countdown, and an isolated run (19 beats) of a rhythm originating adjacent to the atrio-ventricular node with aberrant conduction occurred during the attempted launch of January 27; however, these were not observed at any other time. All of the above are not unexpected physiologic variations. (Samples of MA-6 blockhouse records from the time of insertion and at T- 50 seconds are shown in figs. 9-2 and 9-3.)

Flight

Figure 9-4 illustrates the in-flight physiological data and includes values from the Mercury - Atlas three-orbit centrifuge simulation for comparison. Minute pulse rates were determined by counting every 30 seconds during MA-6 launch and reentry and for 30 seconds at 3 minute intervals throughout the remainder of the flight. Because of the variation in the quality of the respiratory recording, rates were counted for 30 seconds whenever possible and varied from 8 to 19 breaths/minute throughout flight.

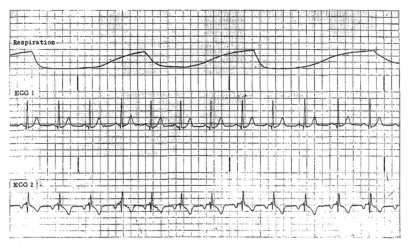

FIGURE 9–2.—Sample of blockhouse physiological record at insertion, 6:06 a.m. e.s.t. Lead 2 is inverted. (Recorder speed 25 mm/sec).

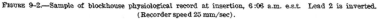

FIGURE 9–3.—Sample of blockhouse record at T–50 seconds, 9:46 a.m. e.s.t., with blood pressure tracing, value 139/80. (Recorder speed 25 mm/sec).

The pulse rate from lift-off to spacecraft separation reached a maximum of 114 beats/minute. The pulse rate varied from 88 to 114 beats/minute in the first 10 minutes of weightlessness. It then remained relatively stable with a mean rate of 86 beats/minute during the next 3 hours and 45 minutes of flight. At the time of retrorocket firing the rate was 96 beats/minute. During reentry acceleration and parachute descent the mean pulse rate was 109 beats/minute, and the highest rate was 134 beats/minute just prior to drogue parachute deployment at a time of maximum spacecraft oscillation. This rate was the highest noted during the mission. These rates indicate that acceleration, weightlessness, and return to gravity were tolerated within acceptable physiological limits. Figure 9-5 compares Astronaut Glenn's in-flight pulse rate, his pulse rate during Mercury - Atlas three-orbit centrifuge simulation, and the mean pulse rate of six astronaut centrifuge simulations. The ECG variations noted during the preflight observation period were not observed in flight. Analysis of the in-flight record revealed only normal sinus rhythm with short periods of sinus bradycardia and sinus arrhythmia. There were rare periods in which trace quality deteriorated so that only pulse rate determinations were possible. The ECG variations noted during Astronaut Glenn's Mercury - Atlas three-orbit centrifuge simulation included: sinus arrhythmia, sinus bradycardia, atrial and nodal premature beats, and rare premature ventricular contractions. These are interpreted as normal physiological variations.

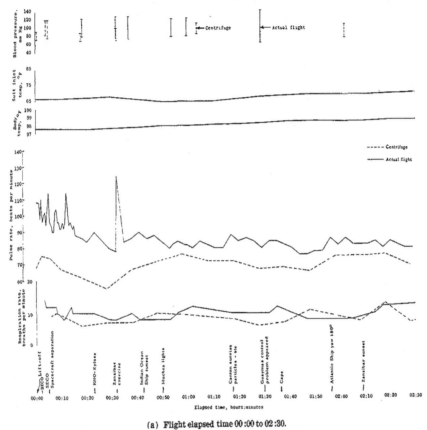

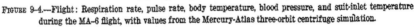

(a) Flight elapsed time 00:00 to 02:30.

FIGURE 9-4.—Flight: Respiration rate, pulse rate, body temperature, blood pressure, and suit-inlet temperature during the MA-6 flight, with values from the Mercury-Atlas three-orbit centrifuge simulation.

Ten blood pressure determinations were made in flight; the first at T + 18 minutes and 30 seconds and the last at f+3 hours and 14 minutes. The values are shown in figure 9-4, and range from 119 to 143 min Hg systolic and from 60 to 81 mm Hg diastolic. The mean blood pressure values and the ranges from physical examinations, static procedures trainer simulations, Mercury-Atlas three orbit centrifuge simulations, launch pad tests, MA-6 countdown, and the MA-6 flight are presented in the following table:

Data sources	Number of determinations	Mean blood pressure, mm Hg	Mean pulse pressure	Systolic Range, mm Hg	Diastolic range, mm Hg
Physical exams	14	110/66	44	98 to 128	60 to 80
Procedures trainer	15	121/76	45	110 to 132	66 to 87
3-orbit Mercury-Atlas centrifuge simulation	56	114/80	34	92 to 136	68 to 92
Launch-pad tests	26	104/76	29	91 to 125	64 to 91
MA-6 countdown	14	123/87	36	101 to 139	83 to 93
MA-6 flight	10	129/70	59	119 to 143	60 to 81

The MA-6 in-flight mean pulse pressure shows some widening when compared with preflight values. This widening appeared after 1 hour of flight and is of uncertain physiological significance. Samples of physiological data from playback of the onboard tape and from range stations are shown in figures 9-4 to 9-9.

The in-flight exercise device is illustrated in figure 9-10. Exercise was accomplished by a series of pulls on elastic bungee cords. An exercise period over Zanzibar on the first orbit raised the pilot's pulse rate from 80 beats/minute to 124 beats/minute after 30 seconds. The pulse rate returned to 84 beats/minute within 2 minutes. The blood pressure before exercise was 129/76 and after exercise was 129/74. This response is within the previously observed values from exercise in the procedures trainer. The environmental control system effectively supported the pilot throughout the mission. It should be noted that body temperature gradually rose from a lift-off value of 97.6° F to 99.5° F at the time of biosensor disconnect. The suit-inlet temperature increased slowly during most of the flight with a more rapid rise after reentry and during parachute descent. During descent and while awaiting recovery on the water, the suit-inlet temperature increased approximately 1° F per minute for a 15-minute period and probably contributed to the pilot's overheated status observed at egress. Since biosensor disconnect occurred 13 minutes before loss of telemetry signal, the maximum body temperature may not have been observed.

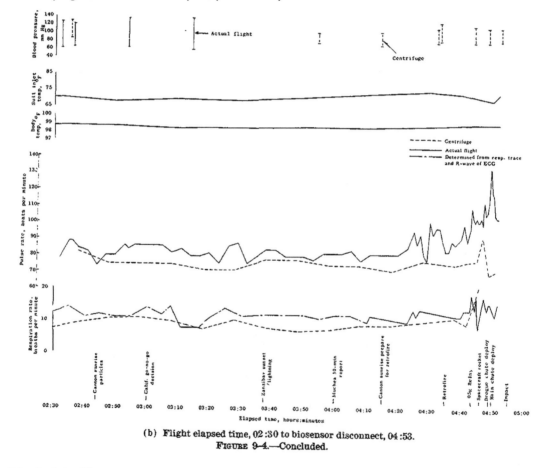

(b) Flight elapsed time, 02:30 to biosensor disconnect, 04:53.
FIGURE 9-4.—Concluded.

Pilot In-flight Observations

The astronaut's voice reports were consistently accurate, confident, and coherent through all phases of the flight. His voice quality conveyed a sense of continued well being and his mental state appeared entirely appropriate for the situation. The pilot's mood and level of performance were effectively conveyed by his voice reports. His prompt responses to ground transmissions and to sounds from the spacecraft suggest no decrement in hearing ability. Visual acuity was maintained and his report of visual perceptions, especially with regard to colors, was accurate and was confirmed by the in-flight photographs.

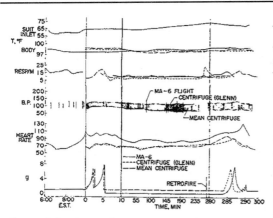

FIGURE 9–5.—Comparison of Astronaut Glenn's inflight physiologic data, his data during the Mercury-Atlas three-orbit centrifuge simulation, and the mean physiologic data of six astronaut centrifuge simulations.

The pilot's voice report contained a number of observations of physiological significance. During his postflight debriefing these reports were amplified. Those considered of most significance are discussed below.

No disturbances in spatial orientation were reported, nor were any symptoms suggestive of vestibular disturbances described during the flight. Voluntary rapid head-turning movements produced no unpleasant sensations. No sensory deprivation or "break-off phenomenon" was noted. A brief sensation of tumbling forward, similar to that described by the astronaut in the MR-4 mission, occurred just after sustainer-engine, cutoff (SECO). This sensation ended promptly and was not associated with nausea. Coincident with retrorocket firing, a feeling of acceleration opposite from flight direction ("back to Hawaii") was noted. This could be expected with the sudden change in

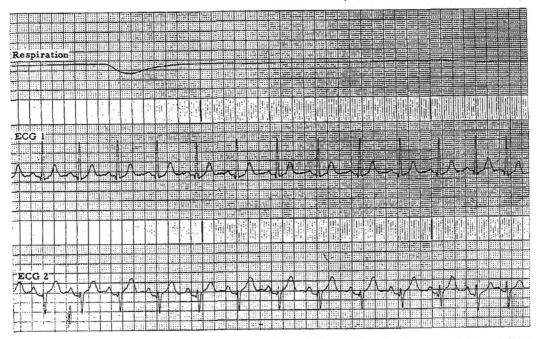

FIGURE 9–6.—Sample of physiological record received at Bermuda Range Station during powered phase of flight, approximately 4 minutes after lift-off. (Recorder speed 25 mm/sec).

spacecraft velocity. The pilot noted no difference, in the sensations associated with reentry accelerations from those experienced during launch.

Food chewing and swallowing were accomplished without difficulty. No water as such was ingested during flight. The pilot urinated without difficulty shortly before reentry. He described "normal" sensations of bladder fullness with the associated urge to urinate.

The astronaut described weightlessness as a "pleasant" sensation and control manipulation was not affected.

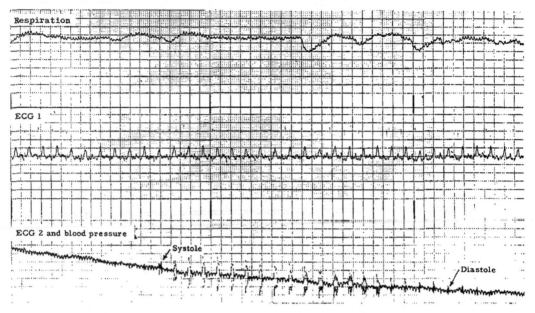

FIGURE 9-7.—Sample of playback record from the onboard tape showing physiological data after 2 hours and 53 minutes of weightlessness, with inflight blood pressure trace, value of 135/64. (Recorder speed 10 mm/sec).

Conclusions

1. The physiological responses observed during the MA-6 mission are all consistent with intact systems and normal body function.

2. The MA-6 mission provided an exposure to weightlessness of sufficient duration to permit physiological responses to reach a relatively steady state.

3. No symptoms reflecting disturbed vestibular function were reported. This lack of findings occurred even though specific attempts were made to stimulate the vestibular system in flight.

4. The pilot's subjective evaluation of his body processes and sensations during the flight all conveyed normal function.

5. Acceleration-weightlessness transition periods did not produce any recognized physiological deterioration. Specifically, reentry acceleration after 4 hours and 38 minutes of weightlessness did not produce any unexpected symptoms and physiological data remained within functional limits.

6. The environmental control system effectively supported the pilot throughout the mission.

Acknowledgments. - The authors gratefully acknowledge the invaluable assistance of : Robie Hackworth, Life Systems Division, NASA Manned Spacecraft Center; and Charles D. Wheelwright, Life Systems Division, NASA Manned Spacecraft Center.

References
1. AUGERSON, WILLIAM S., and LAUGHLIN, C. PATRICK: Physiological Responses of the Astronaut in the MR-3 Flight. Proc. Conf. on Results of the First U.S. Manned Suborbital Space Flight, NASA, Nat Inst. Health, and Nat. Acad. Sci., June 6, 1961, pp. 45—50.
2. LAUGHLIN, C. PATRICK, and AUGERSON, WILLIAM S.: Physiological Responses of the Astronaut in the MR-4 Space Flight. Results of the Second U.S. Manned Suborbital Space Flight, July 21, 1961, NASA Manned Spacecraft Center, pp. 15-21.

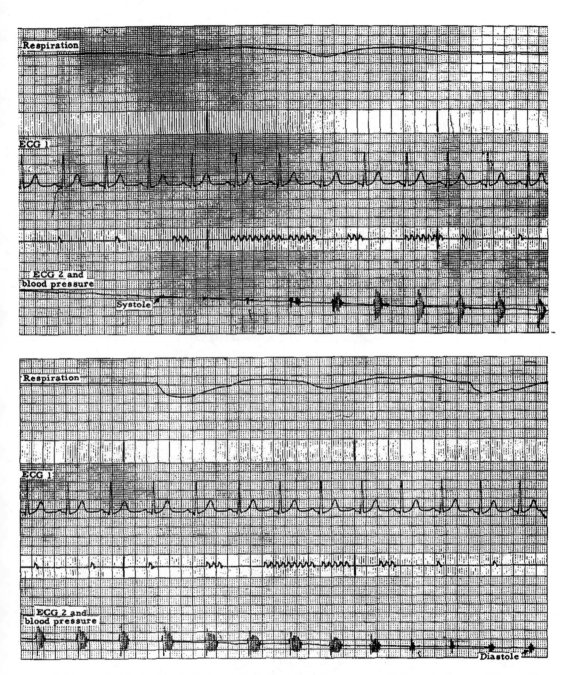

FIGURE 9-8.—Sample of physiological record received at the Hawaii Range Station corresponding to figure 9-7. (Recorder speed 25 mm/sec).

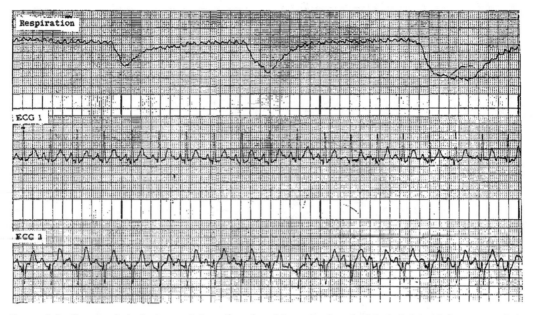

FIGURE 9-9.—Sample of playback record from the onboard tape showing physiological data at drogue parachute deployment, approximately 4 hours and 49 minutes after lift-off.   (Recorder speed 25 mm/sec).

FIGURE 9-10.—MA-6 inflight exercise device.

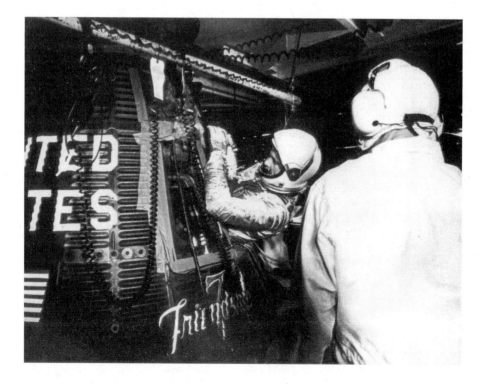

10. ASTRONAUT PREPARATION

By M. SCOTT CARPENTER, Astronaut, NASA Manned Spacecraft Center

Summary

Many hours were profitably spent in specialized training activities, such as spacecraft systems discussions and operation, mission and system procedures and simulated emergencies, physical fitness, and egress and recovery. Also of great value were the many hours the crew spent participating directly in spacecraft preparation and checkout operations. In addition, much time was spent in the study of terrestrial and extraterrestrial features in preparation for scientific and space-navigation observations in orbit. All of these training and study activities contributed greatly to crew readiness for the orbital mission.

Introduction

Since the general Project Mercury training program is common knowledge (see refs. 1 and 2), this discussion is limited to the specialized training activities which were conducted subsequent to the selection of the crew for the MA-6 flight.

Spacecraft Familiarization

At this stage of the Mercury program, each spacecraft differs somewhat from its predecessors and a considerable amount of time must be devoted to the study of these differences. This study was accomplished in part by system briefings conducted by McDonnell Aircraft Corporation and NASA engineers as shown in figure 10-1. Approximately 40 hours were spent in formal briefings of this type. Detailed discussion of environmental control, reaction control, automatic stabilization and control, sequential, electrical, pyrotechnic, communications, and recovery systems were held in the crew quarters by systems engineers and were attended by the flight crew and representatives of the NASA Manned Spacecraft Center Training Division. In addition, many hours of individual study were devoted to the notes and publications which applied specifically to spacecraft 13. (See fig. 10-2.)

FIGURE 10-1.—Spacecraft 13 systems briefings.

A second important activity which contributes measurably to pilot familiarity with the spacecraft is participation, as spacecraft observer, in the many systems checks (see figs. 10-3 and 10-4) which constitute the preparation of the spacecraft for flight. This testing takes place both in the hangar and on the launch pad after mating of the launch-vehicle and spacecraft. A total of over 100 hours was spent in the spacecraft by the flight crew during these tests.

Baseline Physiological Studies

During the early phases of the training, an effort was made to acquire familiarity with the physiological sensations that might be expected during the flight. At the Naval School of Aviation Medicine, Pensacola, Fla., the flight crew received a refresher course in night vision, and spent periods in the slowly revolving room and in the human disorientation device.

Baseline studies of the pilot's individual balance mechanisms were made at this time. (See figs. 10-5 and 10-6). Since an important objective of the flight was to evaluate the astronaut's tolerance of prolonged periods of weightlessness, baseline studies were conducted in an attempt to provide data for comparison with information accumulated during and after the flight. The special equipment, which was developed for in-flight evaluation of orientation ability, is discussed subsequently.

Flight Simulation

The flight crew spent a total of approximately 90 hours in the procedures trainer during which complete mission simulations, both with and without range support, were practiced (See fig. 10-7.) These simulations

FIGURE 10-3.—Astronaut Glenn discusses preparation activities with backup Astronaut Carpenter during spacecraft test of environmental control system.

FIGURE 10-2.—Individual study by Astronauts Glenn and Carpenter.

FIGURE 10-4.—Astronaut Carpenter enters Mercury spacecraft 13 during a preparation test in hanger prior to MA-6 launch.

provided experience in the performance of all flight-plan activities and familiarity with range procedures.

Many hours were spent practicing manual control of spacecraft attitudes. Emphasis was placed on control of the retrofire maneuver turnaround following sustainer engine cutoff (SECO), and reaction control system (RCS) checks following insertion. Orbit maneuvering with low thrusters was also simulated. Additional practice in the manual control task was acquired through the use of the air lubricated free-attitude (ALFA) trainer at Langley Air Force Base, Va. System failures in orbit which required immediate or end of orbit re-entries were practiced and discussed.

The majority of trainer time was devoted to launch aborts with the support of Mercury Control Center (MCC) and Bermuda (BDA) - Astronaut Glenn was subjected to simulated system malfunctions of every description. Some of these, with proper corrective action, resulted in continuation of the mission while others required either immediate or fixed time aborts. These aborts, depending on their nature, could be initiated by either the Astronaut or MCC, or both.

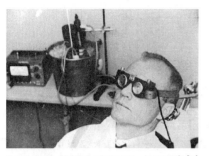

FIGURE 10-5.—Modified caloric test. Astronaut's balance mechanism (semicircular canals) are tested by running cool water into ear and measuring effect on eye motions (nystagmus).

Tape recordings of Astronaut Glenn's voice were made during these trainer sessions and sent to all range stations so that flight controllers might become familiar with his voice and normal manner of speaking. In addition, physiological and performance data were recorded for postflight comparison with onboard data. One additional function of the procedures trainer worthy of note is the opportunity it provides to evaluate the pressure suit in the spacecraft environment. The suit restricts mobility considerably, and procedures as well as the special equipment were designed with this limited mobility in mind. These simulations were excellent not only from the training standpoint, but because they stimulated original thinking that was range wide and many flight-plan and mission-rule inputs resulted. Much was learned both by the astronaut and the flight-control teams. If one activity were to be singled out as being the most valuable in preparing for the flight, it would be this procedures training.

Physical Training

Since its inception, the Mercury physical training program has been the option of the individual. Astronaut Glenn has elected to exercise by running. (See fig. 10-8.) Over the last 3-year period, he has steadily built up from 1 mile to 5 miles a day. For the 3 months preceding the flight he ran 5 miles nearly every day, except for the final week when he tapered off to 2 miles, 1 mile and then 2 days of complete rest prior to the flight.

This activity, including dressing and showering, required about 1 hour per day. It is felt that this is a reasonable amount of time to be so devoted and anything much short of this is insufficient to maintain good physical condition.

FIGURE 10-6.—Ataxia test: Checking Astronaut Glenn's balance mechanism performance by his walking on a narrow board.

FIGURE 10-7.—Astronaut Glenn using procedures trainer for simulated mission.

Special Observation Requirements

A 2-day period was spent at the Morehead Planetarium in Chapel Hill, N.C.. This proved to be an invaluable aid in familiarization with the heavens in general and particularly with those constellations and star patterns that might reasonably be visible through the window for the MA-6 launch date. Members of the Morehead

FIGURE 10-8.—Astronaut Glenn during physical training.

staff were most cooperative and continued use of their facilities is recommended.

Additional study of the constellations was aided by the use of a Farquahr celestial sphere and many star charts, astronomy books, and star finders. A star chart, which proved to be not only a valuable study aid but also a good navigation aid and darkside yaw reference device for in-flight use, was developed.

Two briefings with the Ad Hoc Committee for Astronomical Tasks for the Mercury Astronauts and with scientists of the Project Mercury Weather Support Group and the U.S. Weather Bureau Meteorological Satellite Laboratory were held in Washington, D.C., during which observations of interest to both agencies were discussed. Development of special equipment resulted from these discussions, also, and is covered in the next section.

Special Equipment

As a result of the preflight briefings, the need for some special equipment was apparent, and a container for this equipment was needed. Following are brief descriptions of this equipment.

Two economy-type toothpaste tubes were filled separately with applesauce and beef stew. A screw-on straw which punctured the seal on the top of the tube was provided to duct the food over the lip of the helmet to the mouth.

The pill tube held 10 pills and was spring-loaded for easy extraction of the pills. Each pill measured about three eighths of an inch thick and ¾ inch in diameter. Nine of the pills were chocolate malt tablets and the other was made of xylose which is a five-carbon, traceable form of sugar and was included to measure the rate at which the intestine absorbs food during weightlessness.

Pliers were included to facilitate egress through the top of the spacecraft if the pip pins on the parachute canister became jammed or in the event of a survival situation where pliers have no substitute. The bulb block contained extra amber, green, and red bulbs to be used in the event of telelight or warning-light bulb failure. The waterproof bag was provided for film stowage after landing and before recovery. The camera filter was provided for use with the infrared film and was to be mounted inside the camera when the infrared film was used. Extra film was carried for the regular camera; only the one roll of ultraviolet film already in the ultraviolet camera was carried. The ultraviolet spectrograph consisting of a 36-mm camera equipped with a special quartz lens and prism system was developed for use through the spacecraft window in the 2,000 to 3,000 angstrom wavelength band. A demountable reticle was provided for sighting on the star.

A 35-mm camera with a 50-mm F2.8 lens and a photocell which automatically adjusted the F stop was used for daylight photography. Considerable development effort was required to modify the camera for use in the spacecraft by the astronaut in a pressure suit.

The airglow filter is a device which filters out all light except the 5,577 angstrom wavelength, one of the bright lines of the airglow spectrum. It was intended to be used as an aid in studying the patterning of the airglow layer.

The binoculars were of a miniature type, 8 power with 50-mm objectives.

The filter block was provided for use with the V-Meter. It allowed all the normal exterior observations to be made while excluding all but red, green, or blue light. The V-Meter is a very clever little instrument whose assigned name is the extinctospectropolariscope - occulogyrogravoadaptometer. This device is designed to be used for 16 astronomical and physiological tests. It could be used for measuring the relative brightness of the zodiacal light and other dim night phenomena. It was equipped with crossed polaroid filters which permitted direct viewing of the solar disc and measurement of the polarization of the corona. It could also be used to judge the horizontal under zero-g conditions.

All of this equipment was carried in an accessory kit located by the astronaut's right upper arm. Accessibility was not good but it was the only space available. Use of the equipment was further hampered by the need for a restraining line to each item which was secured to the accessory kit. Velcro, a trade name for an adhering material made up of two types of cloth, one with multiple loops, one with multiple hooks which adhere when pressed together, was used extensively inside the spacecraft for restraining the kit contents during flight. In paper 12, Astronaut Glenn discusses the use of this equipment.

In addition to this equipment in the accessory kit, a knee pad, knife, scissors, survival kit, flashlight, star charts, and an orbital chart book with an overlay of worldwide weather were carried.

A discussion of this equipment is pertinent to the astronaut preparation phase because not only was a great amount of time spent in the development and modification of the equipment but a like period was involved in becoming proficient in its use.

<u>Egress and Recovery Practice</u>

Much time had been spent in egress training prior to the crew selection and little remained to do but polish the procedures.

FIGURE 10-9.—HUS helicopter lifting astronaut from spacecraft, side hatch egress practice.

FIGURE 10-10.—HR2S helicopter preparing to remove astronaut from floating spacecraft, egress practice.

FIGURE 10-11.—U.S. destroyer lifting spacecraft from water with standard boat davits and special lifting beam.

Egress from the small end and side hatch of the spacecraft was practiced with both HUS (see fig. 10-9) and HR2S (see fig. 10-10) helicopters at Langley Air Force Base, Va. Egress was practiced by other members of the astronaut team with a destroyer (see fig. 10-11) out of Norfolk, Va., and they reported no problems.

At Cape Canaveral, Fla., two 3-hour periods were spent with the LARC amphibious vehicle in deep-water familiarization with the life raft and survival equipment. Many equipment and packing modifications resulted from this work.

Pad egress practice was accomplished at Cape Canaveral utilizing the Midas Tower (see fig. 10-12) and the M113 armored vehicle. This practice acquainted all launch complex personnel with the problems related to egress from the spacecraft with the launch vehicle in an unsafe condition.

A form of egress training was conducted at the end of each trainer session by going through the actual sequence of events from parachute deployment to actual egress. This practice helped to smooth the existing procedures as well as to develop new ones.

Miscellaneous

Many other studies were conducted which do not fall into any of the previously mentioned categories. A considerable amount of time was spent on:

(1) Star recognition
(2) Morse code practice
(3) Study of aerial photographs
(4) Study of world charts
(5) Study of Tiros photographs
(6) Study of photographs from previous Mercury flights
(7) Study of mission rules
(8) Study of Atlas systems
(9) Attending briefings
(10) Physical examinations
(11) Correction of minor pressure-suit difficulties

FIGURE 10-12.—Astronaut practicing pad egress with aid of egress tower.

Because of the many delays which preceded the launch of MA-6, it was felt in some quarters that Astronaut Glenn was over-trained. On the contrary, there was easily enough work to fill the available preflight period. During the many delays, he continued to train, modify and practice procedures, and work with and modify the accessory equipment. Training data indicated continued improvement up to the day of launch. The backup astronaut's role throughout was to participate in as much of the training activity as was consistent with the astronaut's need for direct support and the need for an astronaut as spacecraft observer during system tests. Knowing what is involved in this job, it is difficult to envision mission accomplishment in a comparable amount of time, without the services of a backup pilot.

The training period in general went very smoothly. Cooperation was the keynote. A few blind alleys were stumbled into but a sizable extension was made to the trail started by Astronauts Shepard and Grissom. It is hoped that through our efforts, the way for the many who will follow in Astronaut Glenn's footsteps will be a little easier.

References

1. SLAYTON, DONALD K.: Pilot Training and Preflight Preparation, Proc. Conf. on Results of the First U.S. Manned Suborbital Space Flight, NASA, Nat. Inst. Health, and Nat. Acad. Sci., June 6, 1961, pp. 53-60.

2. VOAS, R. B.: Project Mercury: Astronaut Training Program. Physicophysiological Aspects of Space Flight, Columbia Univ. Press, Jan. 1961, pp, 96-116.

11. PILOT PERFORMANCE

By WARREN J. NORTH, Chief, Flight Crew Operations Division, NASA Manned Spacecraft Center; HAROLD I. JOHNSON, Flight Crew Operations Division, NASA Manned Spacecraft Center; HELMUT A. KUEHNEL, Flight Crew Operations Division, NASA Manned Spacecraft Center; and JOHN J. VAN BOCKEL, Flight Crew Operations Division, NASA Manned Spacecraft Center

Summary

The MA-6 flight showed that man can adapt to spacecraft activities in a space environment in much the same way as he adapts to his first flight in a new airplane. The value of static and dynamic simulators in providing accurate spacecraft systems and control familiarization was reaffirmed. Nearly all phases of the MA-6 flight had been simulated. Although Glenn previously experienced zero-gravity flight for durations of only 1 minute in parabolic aircraft flight paths, the extension of weightlessness to 4½ hours caused no concern and was, in fact, a pleasant contrast after spending several hours on his back at 1g.

Proposed concepts for manual control of advanced spacecraft and launch vehicles were given added impetus as a result of the pilot's findings. By giving man a major role in systems operation, as in aircraft practice, the most rapid and efficient attainment of advanced missions will be possible. The possible malfunction of the MA-6 heat-shield release mechanism required the pilot to interrupt the automatic retropackage jettison sequence. The automatic control mode was similarly switched off when the small attitude control jets malfunctioned. The significance of these malfunctions and manual corrective measures can be extrapolated to the design and operational philosophy for highly-complex multistage missions of the future. It is clear that man must play an integral role.

His ability to observe the separated launch vehicle equally well when it was either above or below the horizon, his ability to view the sun safely in space, and his ability to establish a yaw reference optically lends credence to the use of optical rendezvous techniques in Gemini and Apollo missions.

Finally, the performance of Shepard, Grissom, and Glenn during their Mercury flights would seem to justify the selection of mature and experienced aircraft test pilots as Mercury astronauts.

Introduction

The pilot's primary role during the MA-6 flight was to observe and report on spacecraft systems operation and provide control inputs which would insure mission success. Additional activities were included in the flight plan to obtain information on visibility conditions during both day and night, to obtain pictures with special photographic film, and to obtain physiological information. These additional activities were to be conducted only if the spacecraft were operating satisfactorily and did not require full attention from the pilot. In formulating the detailed flight plan, a prime consideration was the orbital position of the spacecraft with respect to the ground tracking and communications stations. It was planned to perform most of the spacecraft maneuvers within line-of-sight distance of the ground stations, in order that spacecraft motions could be correlated between the pilot and the ground readout of UHF radio telemetry. As to be expected on a flight of this nature, radio voice contact was maintained a great majority of the time (approximately 80 percent) .

Although the spacing of Mercury network stations was designed such that the longest void in UHF radio contact would be 17 minutes in a three-orbit mission, during the MA-6 mission the longest gap in effective voice communications was 9 minutes, 13 seconds, apparently due to the longer range of HF radio. This period of radio silence occurred during the third orbit between Australia and Hawaii.

Pilot Performance

Although network communications were excellent during the flight, it was apparent in the Mercury Control

Center that the pilot was the only person with continuous knowledge of spacecraft systems, and he was therefore in the best position to exercise control of the flight. It is significant that even during the period in which he was assessing the control system and the apparent heat-shield malfunctions, he was able to continue detailed systems reporting, make and record visual observations of weather and astronomical phenomena, and take many photographs.

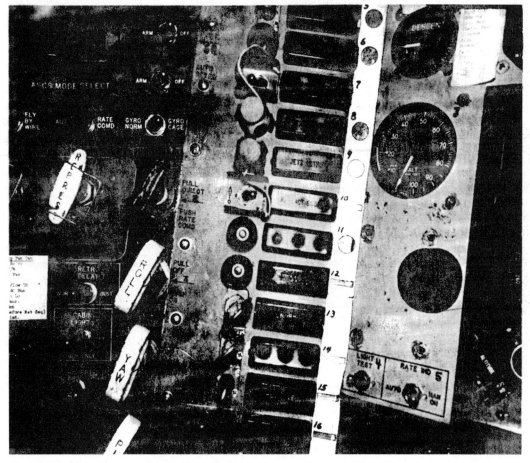

FIGURE 11-1.—MA-6 instrument panel, left section.

One of the important pilot tasks was monitoring the occurrence of critical spacecraft sequential events and providing manual override when necessary. Most of the spacecraft flight events can be identified by instrumentation on the control panel. In this flight, as well as in the previous two manned ballistic flights, the pilot's first and most reliable indication was the actual visual observation and/or an auditory cue of the event and/or a corresponding acceleration from the event, Therefore, the pilot has positive evidence of the occurrence of an event by direct cues without the dependence upon electronic equipment.

Because of the malfunction of the automatic control jets, the pilot was on manual control during most of the last two orbits. The decision to retain the retrorocket package required that the automatic retro jettison switch be left in the "off" position after retrofire. The "off" position of this switch electrically interrupted the sequential system and made it necessary for the pilot to control manually certain events from that time through the end of the flight. Those events were: to retain retro rocket package, to pitch to reentry attitude, to retract the periscope, to actuate 0.05g reentry relay, and to extend the periscope. The rescue aids were deployed manually after impact in accordance with normal procedures.

In each case of manual control of an event, the pilot took the appropriate action and obtained the desired result. The general ability of the pilot to control the vehicle manually is illustrated by a brief review of the major attitude maneuvers accomplished. With the exception of the 180° yaw turnaround, the following maneuvers were required to assess the status of the control system properly and to accomplish the mission:

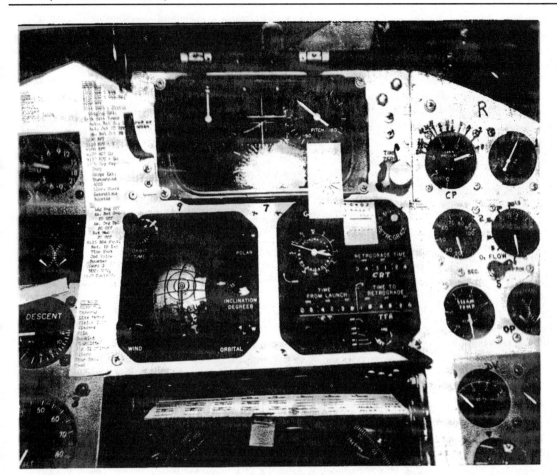

FIGURE 11-2.—MA-6 instrument panel, center section.

Control Systems Check

A control systems check was performed during the first orbit to verify the operational status of the reaction control modes in a minimum amount of time and with minimal fuel usage. Figures 11-1 and 11-2 show the portions of the instrument panel which are involved during the control systems check. The large vertically-oriented handles on the left section are used to control valves in the hydrogen peroxide plumbing system. The horizontal switch array is used to select various electronic control modes. Spacecraft rate and attitude displays are shown in the upper portion of figure 11-2. The control system is designed so that the automatic and manual systems can be used independently or concurrently. Many, additional control combinations are possible in that the manual system can be used to control motions about one axis and the automatic system to command motions about the remaining two axes.

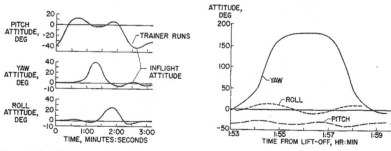

FIGURE 11-3.—MA-6 manual control systems check.

FIGURE 11-4.—MA-6 180° turnaround maneuver.

Figure 11-3 illustrates the in-flight control systems check with a background envelope consisting of four control systems checks accomplished on the procedures trainer. All spacecraft attitudes are defined as zero when the pilot is sitting upright with the small end of the spacecraft pointing horizontally backward along the flight path. As can be seen from this figure, the attitudes between the flight and procedures trainer varied 10° or less; the rates varied less than 1° per second. The time used, to complete the flight maneuver was almost identical to those on the procedures trainer. These in-flight data may give the initial impression of a poorly controlled maneuver; however, it should be remembered that during orbital flight, the spacecraft has no aerodynamic stabilizing forces or aerodynamic damping. Except for retrofire, there is no need to control spacecraft attitudes precisely; consequently, during these maneuvers the pilot was looking for qualitative rather than quantitative results.

<u>180° Right Yaw Maneuver</u>

The pilot made three 180° yaw maneuvers during this flight; however, only the first was intended as a planned maneuver in which the pitch and roll errors were minimized. The other two 180° maneuvers were done for the purpose of observing and taking pictures of the sunrise and the particles surrounding the spacecraft. He had no difficulty making the precise 180° yaw turnaround, using the window, while holding pitch and roll reasonably steady as can be seen in figure 11-4.

These maneuvers involving large deviations from normal spacecraft attitudes produce spurious indications from the horizon scanner gyroscope reference system. In each case, the pilot noted the erroneous indication and re-erected his gyroscopes within the limits required to allow the automatic reference system to restore the proper indication. A similar gyroscope caging operation is necessary to correct aircraft gyro precession after a gross maneuver.

<u>Retrofire Control</u>

The pilot backed up the automatic control system during the retrosequence, and retrofire events, using the manual proportional control mode. The attitudes did not deviate more than 3° during this maneuver. It is difficult to evaluate the effect of the pilot control during retrofire because of concurrent activity of the automatic and manual control jets.

<u>Reentry Pitch Maneuver</u>

He manually positioned the spacecraft to the proper reentry attitude (0° about all axes) using the manual control mode and the rate and attitude instruments. The precise reentry pitch alignment is 1.6°. As can be seen by figure 11-5, he performed this maneuver smoothly and accurately. The initial conditions represent spacecraft attitude at the end of retrofire in which the pitch angle was - 34°.

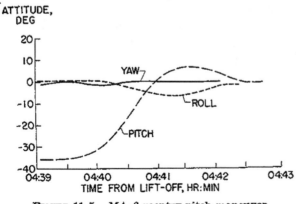

FIGURE 11-5.—MA-6 reentry pitch maneuver.

<u>Reentry Damping</u>
The shape and weight distribution of the Mercury spacecraft provide sufficient inherent aerodynamic stability to reenter with the control system inoperative; in fact, the Mercury "Big Joe" development flight, in September 1959 did reenter successfully with an inactive control system. Consequently, the thrust levels of the reaction-jet control system were designed with sufficient capability to control spacecraft attitudes during retrofire, but not to control spacecraft trim attitudes during the high aerodynamic forces of reentry. The pilot's reentry task was to align the spacecraft

initially with the velocity vector, then to damp or minimize the low-frequency oscillations which occur. Although his attention was diverted by the burning retropack, reentry damping was performed satisfactorily by using a combination of the manual and automatic control systems until the control fuel was depleted.

In contrast to the MA-5 mission, there was little concern regarding the ability of the manned MA-6 spacecraft to complete three orbits. In fact, the third orbit gave the pilot additional time to experiment with the mal-functioning control system in order that he could better perform a successful retrofire and reentry.

Pilot Observations

In general, the pilot found he could easily orient the spacecraft in pitch and roll by using the external horizon reference.

Yaw (or drift) is more difficult to determine at orbital attitudes because ground terrain features and clouds subtend low angular rates with respect to the spacecraft. By the end of the flight, he was able to determine yaw quite easily, both on the daylight side and during full moonlight night conditions, by using the window reference. During the flight, he used the procedure of pitching down to -60° to pick up terrain drift due to the orbital velocity. He found that the periscope was not as useful as the window for determining drift on the nightside. Even with a full moon, the clouds were too dim in the periscope to pick up readily a specific point and follow it for yaw heading information.

The pilot was able to observe the separated launch vehicle clearly when it was both above and below the horizon. Direct observation of the sun through the window was no more annoying than direct observations from the surface of the earth.

John Glenn describes these and other observations fully in paper 12.

He found that weightlessness was pleasant, and in several respects, easier or more enjoyable than the 1g condition. Zero-g facilitated certain tasks, such as using the camera, since this equipment could be left hanging in midair when he was interrupted by other activities. The pilot experienced no problem in reaching for and activating controls. The effects of head rotation in a zero-g field were investigated. He moved his head rapidly in each of the three planes of rotation, with no sensations of nausea or vertigo. The pilot reported that he could feel only the highest angular accelerations encountered during the flight. Most of the attitude maneuvers were conducted at rates lower than those that could be sensed under 1g.

12. PILOT'S FLIGHT REPORT

BY JOHN H. GLENN, Jr., Astronaut, NASA Manned Spacecraft Center

Summary

Weightless flight was quickly adapted to, and was found to be pleasant and without discomfort. The chances of mission success are greatly enhanced by the presence of a human crew in the spacecraft. A human crew is vital to future space missions for the purpose of intelligent observation and actions when the spacecraft encounters expected or unexpected occurrences or phenomena.

Introduction

The test objectives for the MA-6 mission of Friendship 7, as quoted from the Mission Directive, were as follows:
(1) Evaluate the performance of a man/spacecraft system in a three-orbit mission

(2) Evaluate the effects of space flight on the astronaut

(3) Obtain the astronaut's opinions on the operational suitability of the spacecraft and supporting systems for manned space flight.

These are obviously broad objectives. Previous papers have described in some detail the operation of the spacecraft systems and, to a degree, man's integration with these systems.

My report is concerned mainly with those items in all three objectives where man's observation capabilities provided information not attained by other means. It is in this type of reporting that a manned vehicle provides a great advantage over an unmanned vehicle, which is often deaf and blind to the new and the unexpected. My report, then, will stress what I heard, saw, and felt during the orbital flight.

Preparation and Countdown

Preparation, transfer to the launch pad, and insertion into the spacecraft went as planned,

The technicians and I had been through the entry to the spacecraft many times. As with every countdown, short delays were encountered when problems arose. The support for the microphone in the helmet, an item that had been moved and adjusted literally thousands of times, broke and had to be replaced. While the spacecraft hatch was being secured, a bolt was broken and had to be repaired. During this time I was busy going over my checklist and monitoring the spacecraft instruments. Many people were concerned about my mental state during this and earlier delays, which are a part of preparation for a manned space flight. People have repeatedly asked whether I was afraid before the mission. Humans always have fear of an unknown situation, this is normal. The important thing is what we do about it. If fear is permitted to become a paralyzing thing that interferes with proper action, then it is harmful. The best antidote to fear is to know all we can about a situation. It is lack of knowledge which often misleads people when they try to imagine the feelings of an astronaut about to launch. During the years of preparation for Project Mercury, the unknown areas have been shrunk, we feel, to an acceptable level. For those who have not had the advantage of this training, the unknowns appear huge and insurmountable, and the level of confidence of the uninformed is lowered by an appropriate amount.

All the members of the Mercury team have been working towards this space flight opportunity for a long time. We have not dreaded it; we have looked forward to it. After 3 years we cannot be unduly concerned by a few delays. The important consideration is that everything be ready, that nothing be jeopardized by haste which can be preserved by prudent action.

The initial unusual experience of the mission is that of being on top of the Atlas launch vehicle after the gantry has been pulled back. Through the periscope, much of Cape Canaveral can be seen. If you move back and forth in the couch, you can feel the entire vehicle moving very slightly. When the engines are gimbaled, you can feel the vibration. When the tank is filled with liquid oxygen, the spacecraft vibrates and shudders as the metal skin flexes. Through the window and periscope the white plume of the lox (liquid oxygen) venting is visible.

Launch

When the countdown reached zero, I could feel the engines start. The spacecraft shook, not violently but very solidly. There was no doubt when lift-off occurred. When the Atlas was released there was an immediate gentle surge that let you know you were on your way. The roll to correct azimuth was noticeable after lift-off. I had preset the little window mirror to watch the ground. I glanced up after lift-off and could see the horizon turning. Some vibration occurred immediately after lift- off. It smoothed out after about 10 to 15 seconds of flight but never completely stopped. There was still a noticeable amount of vibration that continued up to the time the spacecraft passed through the maximum aerodynamic pressure or maximum q, at approximately T+1 minute. The approach of maximum q is signaled by more intense vibrations. Force on the outside of the spacecraft was calculated at 982 pounds per square foot at this time, During this period, I was conscious of a dull muffled roar from the engines. Beyond the high q area the vibration smoothed out noticeably. However, the spacecraft never became completely vibration free during powered flight.

The acceleration buildup was noticeable but not bothersome. Before the flight my backup pilot, Astronaut Scott Carpenter, had said he thought it would feel good to go in a straight-line acceleration rather than just in circles as we had in the centrifuge and he was right. Booster engine cut-off occurred at 2 minutes 9.6 seconds after lift-off. As the two outboard engines shut down and were detached, the acceleration dropped but not as sharply as I had anticipated. Instead, it decayed over approximately ½ second. There is a change in noise level and vibration when these engines are jettisoned. I saw a flash of smoke out the window and thought at first that the escape tower had jettisoned early and so reported. However, this flash was apparently deflected smoke coming up around the spacecraft from the booster engines which had just separated. The tower was jettisoned at 2 minutes, 33.3 seconds, and I corrected my earlier report. I was ready to back up the automatic sequencing system if it did not perform correctly and counted down the seconds to the time for tower jettisoning. I was looking at the nozzles of the tower rockets when they fired. A large cloud of smoke came out but little flame. The tower accelerated rapidly from the spacecraft in a straight line. I watched it to a distance of approximately ½ mile. The spacecraft was programmed to pitch down slowly just prior to jettisoning the tower and this maneuver provided my first real view of the horizon and clouds. I could just see clouds and the horizon behind the tower as it jettisoned.

After the tower fired, the spacecraft pitched slowly up again and I lost sight of the horizon. I remember making a comment at about this time that the sky was very black. The acceleration built up again, but as before, acceleration was not a major problem. I could communicate well, up to the maximum of 7.7g at insertion when the sustainer-engine thrust terminates. Just before the end of powered flight, there was one experience I was not expecting. At this time the fuel and lox tanks were getting empty and apparently the Atlas becomes considerably more flexible than when filled. I had the sensation of being out on the end of a spring board and could feel oscillating motions as if the nose of the launch vehicle were waving back and forth slightly. (Appendix B presents the onboard tape transcript of the Friendship 7 orbital flight.)

Insertion into Orbit

The noise also increased as the vehicle approached SECO (sustainer engine cutoff). When the sustainer engine cutoff at 5 minutes, 1.4 seconds and the acceleration dropped to zero, I had a slight sensation of tumbling forward. The astronauts have often had a similar sensation during training on the centrifuge. The sensation, was much less during the flight, and since the spacecraft did pitch down at this point it may have been a result of actual movement rather than an illusion.

There was no doubt when the clamp ring between the Atlas and the Mercury spacecraft fired. There was a loud report and I immediately felt the force of the posigrade rockets which separate the spacecraft from the launch vehicle. Prior to the flight I had imagined that the acceleration from these three small rockets would be insignificant and that we might fail to sense them entirely, but there is no doubt when they fire.

Immediately after separation from the Atlas, the autopilot started to turn the spacecraft around. As the spacecraft came around to its normal aft viewing attitude, I could see the Atlas through the window. At the time I estimated that it was "a couple of hundred yards away." After the flight an analysis of the trajectory data showed that the distance between the launch vehicle and the spacecraft should, at this point, be 600 feet. Close enough for a rough estimate. I do not claim that I can normally judge distance so close. There was a large sized luck factor in the estimate; nevertheless, the facts do give an indication that man can make an adequate judgment at least of short distances to a known object in space. This capability will be important in future missions in which man will want to achieve rendezvous, since the pilot will be counted on to perform the final closing maneuver.

I was able to keep the Atlas in sight for 6 or 7 minutes while it traveled across the Atlantic. The last time I reported seeing it the Atlas was approximately 2 miles behind and I mile below the spacecraft. It could be seen easily as a bright object against the black background of space and later against the background of earth.

Orbit

The autopilot turned the spacecraft around and put it into the proper attitude. After my initial contact with Bermuda I received the times for firing the retrorockets and started the check of the controls. This is a test of the control systems aboard the spacecraft. I had practiced it many times on the ground in the Mercury procedures trainer and the test went just as it had in the trainer. I was elated by the precision with which the test progressed. It is quite an intricate check. With your right hand you move the control stick, operating the hydrogen peroxide thrusters to move the spacecraft in roll, pitch, and yaw. With your left hand you switch from one control system to another as the spacecraft is manually controlled to a number of precise rates and attitudes. This experience was the first time I had been in complete manual control, and it was very reassuring to see not only the spacecraft react as expected, but also to see that my own ability to control was as we had hoped. Following this controls check I went back to autopilot control and the spacecraft operated properly on autopilot throughout the first orbit

Thruster Problem

Because of a malfunction in a low-torque thruster at the end of the first orbit, it was necessary to control the spacecraft manually for the last two orbits. This requirement introduced no serious problems, and actually provided me with an opportunity to demonstrate what a man can do in controlling a spacecraft. However, it limited the time that could be spent on many of the experiments I had hoped to carry out during the flight.

Flight Plan

The Mercury flight plan during the first orbit was to maintain optimum spacecraft attitude for radar tracking and communication checks. This plan would provide good trajectory information as early as possible and would give me a chance to adapt to these new conditions if such was necessary. Other observations and tasks were to be accomplished mainly on the second and third orbits. Since the thruster problem made it necessary for me to control manually during most of the second and third orbits, several of the planned observations and experiments were not accomplished.

Attitude Reference

A number of questions have been raised over the ability of a man to use the earth's horizon as a reference for controlling the attitude of the space vehicle. Throughout this flight no trouble in seeing the horizon was encountered. During the day the earth is bright and the background of space is dark. The horizon is vividly marked. At night, before the moon is up, the horizon can still be seen against the background of stars. After the moon rises (during this flight the moon was full), the earth is well enough lighted so that the horizon can be clearly seen.

With this horizon as a reference, the pitch and roll attitudes of the spacecraft can easily be controlled. The window can be positioned where you want it. Yaw, or heading reference, however, is not so good. I believe that there was a learning period during the flight regarding my ability to determine yaw. Use of the view through the window and periscope gradually improved. To determine yaw in the spacecraft, advantage must be taken of the speed of the spacecraft over the earth which produces an apparent drift of the ground below the spacecraft. When the spacecraft is properly oriented, facing along the plane of the orbit, the ground appears to move parallel to the spacecraft longitudinal axis. During the flight I developed a procedure which seemed to help me use this terrain drift as a yaw reference. I would pitch the small end of the spacecraft down to about -60° from the normal attitude where a fairly good vertical view was available. In this attitude, clouds and land moving out from under me had more apparent motion than when the spacecraft was in its normal orbit attitude and I looked off toward the horizon.

At night with the full moon illuminating the clouds below, I could still determine yaw through the window but not as rapidly as in the daytime. At night I could also use the drift of the stars to determine heading but this took longer and was less accurate. Throughout the flight I preferred the window to the periscope as an attitude reference system. It seemed to take longer to adjust yaw by using the periscope on the day side. At night, the cloud illumination by the moon is too dim to be seen well through the periscope.

Three times during the flight I turned the spacecraft approximately 180° in yaw and faced forward in the direction of flight. I liked this attitude - seeing where I was going rather than where I had been - much better. As a result of these maneuvers my instrument reference system gave me an inaccurate attitude indication. It was easy to determine the proper attitude, however, from reference to the horizon through the window or to the periscope. Maintaining orientation was no problem, but I believe that the pilot automatically relies much more completely on vision in space than he does in an airplane, where gravity cues are available. The success with which I was able to control the spacecraft at all times was, to me, one of the most significant features of the flight.

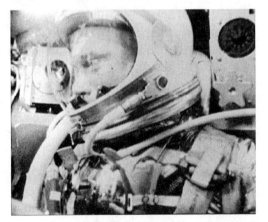

FIGURE 12-1.—Pilot looks to his right. Note the distance his head can be turned in the pressure suit.

Weightlessness

Weightlessness was a pleasant experience. I reported I felt fine as soon as the spacecraft separated from the launch vehicle, and throughout the flight this feeling continued to be the same.

Approximately every 30 minutes, throughout the flight I went through a series of exercises to determine whether weightlessness was affecting me in any way. To see if head movement in a zero g environment produced any symptoms of nausea or vertigo, I tried first moving, then shaking my head from side to side, up and down, and tilting it from shoulder to shoulder. In other words, moving my head in roll, pitch, and yaw. I began slowly, but as the flight progressed, I moved my head more rapidly and vigorously until at the end of the flight I was moving as rapidly as my pressure suit would allow. In figure 12-1[1] the camera has caught me in the middle of this test, and this photograph shows the extent to which I was moving my head.

In another test, using only eye motions, I tracked a rapidly moving spot of light generated by my finger-tip lights. I had no problem watching the spot and once again no sensations of dizziness or nausea. A small eye chart was included on the instrument panel, with letters of varying size and with a "spoked wheel" pattern to check both general vision and any tendency toward astigmatism. No change from normal was apparent.

An "oculogyric test" was made in which turning rates of the spacecraft were correlated with sensations and

[1] All the originals of these photographs are in color and some detail is lost in the black and white reproduction of these photographs .

eye movements. Results were normal. Preflight experience in this test and a calibration had been made at the Naval School of Aviation Medicine, Pensacola, Fla., with Dr. Ashton Graybiel, so that I was thoroughly familiar with my reactions to these same movements at 1 g.

To provide medical data on the cardiovascular system, at intervals, I did an exercise which consisted of pulling on a bungee cord once a second for 90 seconds. This exercise provided a known workload to compare with previous similar tests made on the ground. The flight surgeons have reported the effect that this had on my pulse and blood pressure. The effect that it had on me during the flight was the same effect that is had on the ground — it made me tired.

Another experiment related to the possible medical effects of weightlessness was eating in orbit. (See fig. 12-2.) On the relatively short flight of Friendship 7, eating was not a necessity, but rather an attempt to determine whether there would be any problem in consuming and digesting food in a weightless state. At no time did I have any difficulty eating. I believe that any type of food can be eaten as long as it does not come apart easily or make crumbs. Prior to the flight, we joked about taking along some normal food such as a ham sandwich. I think this would be practical and should be tried.

FIGURE 12-2.—Pilot opens visor to eat.

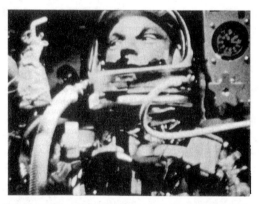

FIGURE 12-3.—After his snack of applesauce, the pilot leaves his expended tube hanging in air momentarily.

Sitting in the spacecraft under zero g is more pleasant than under 1 g on the ground, since you are not subject to any pressure points. I felt that I adapted very rapidly to weightlessness. I had no tendency to overreach nor did I experience any other sign of lack of coordination, even on the first movements after separation. I found myself unconsciously taking advantage of the weightless condition, as when I would leave a camera or some other object floating in space while I attended to other matters. This was not done as a preplanned maneuver but as a spur-of-the-moment thing when another system needed my attention. I thought later about how I had done this as naturally as if I were laying the camera on a table in a 1 g field. It pointedly illustrates how rapidly adaptable the human is, even to something as foreign as weightlessness. (See fig. 12-3.) We discovered from this flight that some problems are still to be solved in properly determining how to stow and secure equipment that is used in a space vehicle. I had brought along a number of instruments, such as, cameras, binoculars, and a photometer, with which to make observations from the spacecraft. All of these were stowed in a ditty bag by my right arm. Each piece of equipment had a 3-foot piece of line attached to it. By the time I had started using items of the equipment, these lines became tangled. Although these lines got in the way, it was still important to have some way of securing the equipment, as I found out when I attempted to change film. The small canisters of film were not tied to the ditty bag by line & I left one floating in midair while working with the camera, and when I reached for it, I accidentally hit it and it floated out of sight behind the instrument panel

Color and Light [2]

As I looked back at the earth from space, colors and light intensities were much the same as I had observed when flying at high altitude in an airplane. The colors observed when looking down at the ground appeared

[2] A more detailed description of the visual observations taken from the postflight debriefing is presented in the appendix.

similar to those seen from 50,000 feet, When looking toward the horizon, however, the view is completely different, for then the blackness of space contrasts vividly with the brightness of the earth. The horizon itself is a brilliant, brilliant blue and white.

It was surprising how much of the earth's surface was covered by clouds. The clouds can be seen very clearly on the daylight side. The different types of clouds — vertical developments, stratus clouds, and cumulus clouds — are readily distinguished. There is little problem identifying them or in seeing the weather patterns. You can estimate the relative heights of the cloud layers from your knowledge of the types or from the shadows the high clouds cast on those lower down. These observations are representative of information which the scientists of the U.S. Weather Bureau Meteorological Satellite Laboratory had asked Project Mercury to determine. They are interested in improving the optical equipment in their Tiros and Nimbus satellites and would like to know if they could determine the altitude of cloud layers with better optical resolution, From my flight I would say it is quite possible to determine cloud heights from this orbital altitude. (See figs. 12-4 to 12-8.)

Only a few land areas were visible during the flight because of the cloud cover. Clouds were over much of the Atlantic, but the western (Sahara Desert) part of Africa was clear. As I passed over it the first time I took the picture shown in figure 12-9. In this desert region I could plainly see dust storms. By the time I got to the east coast of Africa where I might have been able to see towns, the land was covered by clouds. The Indian Ocean was the same.

Western Australia was clear, but the eastern half was overcast. Most of the area across Mexico and nearly to New Orleans was covered with high cirrus clouds. As I came across the United States I could see New Orleans, Charleston, and Savannah very clearly. I could also see rivers and lakes. I think the best view I had of any land area during the flight was the clear desert region around El Paso on the second pass across the United States. I could see the colors of the desert and the irrigated area north of El Paso. As I passed off the east coast of the United States I could see across Florida and far back along the Gulf Coast. (See figs. 12-10 and 12-11.)

FIGURE 12-4.—Photograph taken over the Pacific Ocean at the end of the first orbit. The cloud panorama illustrates the visibility of different cloud types and weather patterns. Shadows produced by the rising sun aid in the determination of relative cloud heights.

Over the Atlantic I saw what I assume was the Gulf Stream. The different colors of the water are clearly visible. I also observed what was probably the wake of a ship. As I was passing over the recovery area at the end of the second orbit, I looked down at the water and saw a little "V." I checked the map. I was over recovery area G at the time, so I think it was probably the wake from a recovery ship. When I looked again the little "V" was under a cloud. The change in light reflections caused by the wake of a ship are sometimes visible for long distances from an airplane and will linger for miles behind a ship. This wake was probably what was visible.

I believe, however, that most people have an erroneous conception that from orbital altitude, little detail can be seen. In clear desert air, it is common to see a mountain range 100 or so miles away very clearly, and all

Figure 12-5.—Just before sunset on the first orbit, the pilot's camera catches the darkening earth. The photograph shows how the shadows help to indicate cloud heights.

Figure 12-6.—Over the Atlantic on the third orbit, the pilot's camera shows an overcast region to the northwest and patterns of scattered clouds in the foreground.

that vision is through atmosphere. From orbital altitude, atmospheric light attenuation is only through approximately 100,000 feet of atmosphere so it is even more clear. An interesting experiment for future flights can be to determine visibility of objects of different sizes, colors, and shapes.

Obviously, on the night side of the earth, much less was visible. This may have been due not only to the reduced light, but also partly to the fact that I was never fully dark adapted. In the bright light of the full moon, the clouds are visible. I could see vertical development at night. Most of the cloudy areas, however, appeared to be stratoform.

The lights of the city of Perth, in Western Australia, were on and I could see them well. The view was similar to that seen when flying at high altitude at night over a small town. South of Perth there was a small group of lights, but they were much brighter in intensity. Inland there was a series of four or five towns lying in a line running from east to west. Knowing that Perth was on the coast, I was just barely able to see the coastline of Australia. Clouds covered the area of eastern Australia around Woomera, and I saw nothing but clouds from there across the Pacific until I was east of Hawaii. There appeared to be almost solid cloud cover all the way.

Just off the east coast of Africa were two large storm areas. Weather Bureau scientists had wondered whether lightning could be seen on the night side, and it certainly can. A large storm was visible just north of my track over the Indian Ocean and a smaller one to the south. Lightning could be seen flashing back and forth between the clouds but most prominent were lightning flashes within thunderheads illuminating them like light bulbs.

Some of the most spectacular sights during the flight were sunsets. The sunsets always occurred slightly to my left, and I turned the spacecraft to get a better view. The sunlight coming in the window was very brilliant, with an intense clear white light that reminded me of the arc lights while the spacecraft was on the launching pad.

I watched the first sunset through the photometer (fig. 12-12) which had a polarizing filter on the front so that the intensity of the sun could be reduced to a comfortable level for viewing. Later I found that by squinting, I could look directly at the sun with no ill effects, just as I can from the surface of the earth. This accomplished little of value but does give an idea of intensity.

The sun is perfectly round as it approaches the horizon. It retains most of its symmetry until just the last sliver is visible. The horizon on each side of the sun is extremely bright, and when the sun has gone down to the level of this bright band of the horizon, it seems to spread out to each side of the point where it is setting. With the camera I caught the flattening of the sun just before it set (fig. 12-13 (b)). This is a phenomenon of some interest to the astronomers.

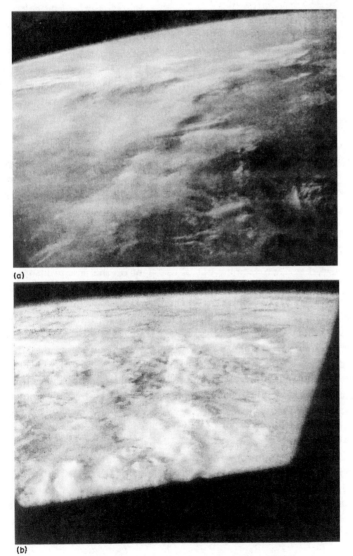

(a)

(b)

As the sun moves toward the horizon, a black shadow of darkness moves across the earth until the whole surface, except for the bright band at the horizon, is dark. This band is extremely bright just as the sun sets, but as time passes the bottom layer becomes a bright orange and fades into reds, then on into the darker colors, and finally off into the blues and blacks. One thing that surprised me was the distance the light extends on the horizon on each side of the point of the sunset. The series of pictures shown in figures 12-13 and 12-14 illustrates the sequence of this orbital twilight. I think that the eye can see a little more of the sunset color band than the camera captures. One point of interest was the length of time during which the orbital twilight persisted. Light was visible along the horizon for 4 to 5 minutes after the sunset, a long time when you consider that sunset occurred 18 times faster than normal.

(a) First orbit.
(b) Second orbit.
FIGURE 12-8.—The Western Indian Ocean was overcast on the first and second orbits. The relative heights of the cirrus and the cumulus clouds can clearly be seen.

The period immediately following sunset was of special interest to the astronomers. Because of atmospheric light scattering, it is not possible to study the region close to the sun except at the time of a solar eclipse. It had been hoped that from above the atmosphere the area close to the sun could be observed. However, this would require a period of dark adaptation prior to sunset. An eye patch had been developed for this purpose, which was to be held in place by special tape. This patch was expected to permit one eye to be night adapted prior to sunset. Unfortunately, the tape proved unsatisfactory and I could not use the eye patch. Observations of the sun's corona and zodiacal light must await future flights when the pilot may have an opportunity to get more fully dark adapted prior to sunset.

FIGURE 12-9.—View looking back toward the African coast on the first orbit. The photograph from the pilot's camera shows the desert with blowing sand in the foreground.

Another experiment suggested by our advisors in astronomy was to obtain ultraviolet spectrographs of the stars in the belt and sword of Orion. The ozone layer of the earth's atmosphere will not pass ultraviolet light below 3,000 angstroms. The spacecraft window will pass light down to 2,000 angstroms. It is possible, therefore, to get pictures of the stars from the Mercury spacecraft which cannot be duplicated by the largest telescopes on the ground. Several ultraviolet spectrographs were taken of the stars in the belt of Orion. They are being studied at the present time to see whether useful information was obtained.

The biggest surprise of the flight occurred at dawn. Coming out of the night on the first orbit, at the first glint of sunlight on the spacecraft, I was looking inside the spacecraft checking instruments for perhaps 15 to 20 seconds. When I glanced back through the window my initial reaction was that the spacecraft had tumbled and that I could see nothing but stars through the window. I realized, however, that I was still in the normal attitude. The spacecraft was surrounded by luminous particles. These particles were a light yellowish green color. It was as if the spacecraft were moving through a field of fireflies. They were about the brightness of a first magnitude star and appeared to vary in size from a pinhead up to possibly 3/8 inch. They were about 8 to 10 feet apart and evenly distributed through the space around the spacecraft. Occasionally one or two of them would move slowly up around the spacecraft and across the window, drifting very, very slowly, and would then gradually move off, back in the direction I was looking. I observed these luminous objects for approximately 4 minutes each time the sun came up.

During the third sunrise I turned the spacecraft around and faced forward to see if I could determine where the particles were coming from. Facing forwards I could see only about 10 percent as many particles as I had when my back was to the sun. Still, they seemed to be coming towards me from some distance so that they appeared not to be coming from the spacecraft. Just what these particles are is still subject to debate and awaits further clarification. Dr. John O' Keefe at the NASA Goddard Space Flight Center is making a study in an attempt to determine what these particles might be. (See appendix D.)

Other Planned Observations

    As mentioned earlier, a number of other observations and measurements during orbit had to be canceled because of the control system problems. Equipment carried was not highly sophisticated scientific equipment. We believed, however, that it would show the feasibility of making more comprehensive measurements on later missions.

Some of these areas of investigation that we planned but did not have an opportunity to check are as follows:

(A) Weather Bureau
    observations:
(1) Pictures of weather areas and cloud formations to match against map forecasts and Tiros pictures
(2) Filter mosaic pictures of major weather centers
(3) Observation of green air glow from air and weather centers in 5,577 angstrom band with air glow filter
(4) Albedo intensities measure reflected light intensities on both day and night side

(B) Astronomical observations:
(1) Light polarization from area of sun
(2) Comets close to sun
(3) Zodiacal light
(4) Sunlight intensity
(5) Lunar clouds
(6) Gegenschein
(7) Starlight intensity measurements

(C) Test for otolith balance disturbance and autokynesis phenomena:

(D) Vision tests:
(1) Night vision adaptation
(2) Phorometer eye measurements

(E) Drinking

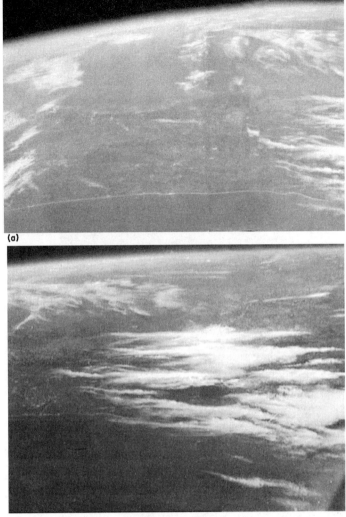

(a)

(b)

FIGURE 12–10.—At the beginning of the third orbit, the pilot catches a panoramic view of the Florida coast, from the cloud covered Georgia border to just above Cape Canaveral.

Reentry

After having turned around on the last orbit to see the particles, I maneuvered into the correct attitude for firing the retrorockets and stowed the equipment in the ditty bag. This last dawn found my attitude indicators

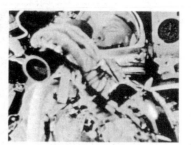

FIGURE 12-11.—View of the Florida area from Tiros IV taken at 1610 G.c.t. on February 20, 1962. This photograph shows the band of clouds (across Southern Florida) which had moved away from Cape Canaveral earlier that morning. The clouds just north of Florida are apparently the ones plainly visible in figure 12-10. (U.S. Weather Bureau photograph; major land masses are outlined in white ink.)

FIGURE 12-12.—During sunset, the pilot used the photometer to view the sun.

still slightly in error. However, before it was time to fire the retrorockets the horizon-scanner slaving mechanism had brought the gyros back to orbit attitude. I cross-checked repeatedly between the instruments, periscope presentation, and the attitude through the window.

Although there were variations in the instrument presentations during the flight, there was never any difficulty in determining my true attitude by reference to the window or periscope, I received a countdown from the ground and the retrorockets were fired on schedule just off the California coast.

I could hear each rocket fire and could feel the surge as the rockets slowed the spacecraft. Coming out of zero-g condition, the retrorocket firing produced the sensation that I was accelerating back toward Hawaii (fig. 12-15). This sensation, of course, was an illusion. Following retrofire the decision was made to have me reenter with the retro package still on because of the uncertainty as to whether the landing bag had been extended. This decision required me to perform manually a number of the operations which are normally automatically programmed during the reentry.

These maneuvers I accomplished. I brought the spacecraft to the proper attitude for reentry under manual control. The periscope was retracted by pumping the manual retraction lever.

As deceleration began to increase I could hear a hissing noise that sounded like small particles brushing against the spacecraft.

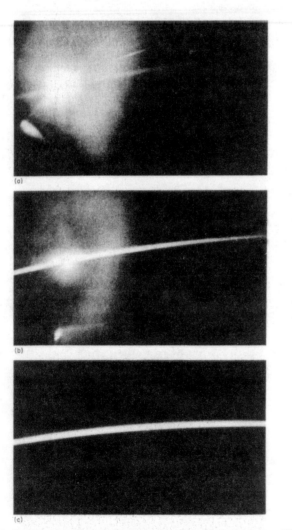

(a)

(b)

(c)

FIGURE 12-13.—The third orbital sunset as shown by a series of three photographs. The camera catches the flattening of the solar disk just before the sun disappears below the horizon.

Due to ionization around the spacecraft, communications were lost. This had occurred on earlier missions and was experienced now on the predicted schedule. As the heat pulse started there was a noise and a bump on the spacecraft. I saw one of the straps that holds the retrorocket package swing in front of the window.

The heat pulse increased until I could see a glowing orange color through the window. Flaming pieces were breaking off and flying past the spacecraft window. (See fig 12-16.) At the time, these observations were of some concern to me because I was not sure what they were. I had assumed that the retropack had been jettisoned when I saw the strap in front of the window. I thought these flaming pieces might be parts of the heat shield breaking off. We know now, of course, that the pieces were from the retropack.

There was no doubt when the heat pulse occurred during reentry but it takes time for the heat to soak into the spacecraft and heat the air. I did not feel particularly hot until we were getting down to about 75,000 to 80,000 feet. From there on down I was uncomfortably warm, and by the time the main parachute was out I was perspiring profusely.

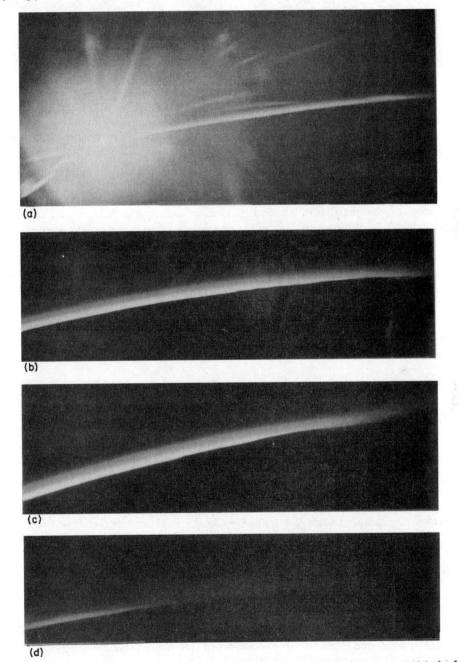

FIGURE 12-14.—The first orbital sunset is recorded in a series of four photographs. Patches of high clouds glow orange as the sun's light is diffracted by the atmosphere. The bright band of space twilight grows dimmer and shrinks in length with passing time.

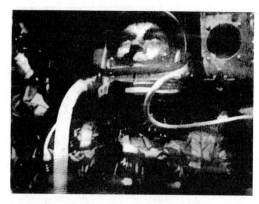

FIGURE 12–15.—Pilot concentrates on instruments while controlling attitude during retrofire.

FIGURE 12–16.—Pilot looks out of window at fireball during maximum reentry heating.

The reentry deceleration of 7.7g was as expected and was similar to that experienced in centrifuge runs. There had been some question as to whether our ability to tolerate acceleration might be worse because of the 4½ hours of weightlessness, but I could note no difference between my feeling of deceleration on this flight and my training sessions in the centrifuge.

After peak deceleration, the amplitude of the spacecraft oscillations began to build. I kept them under control on the manual and fly-by-wire systems until I ran out of manual fuel. After that point, I was unknowingly left with only the fly-by-wire system and the oscillations increased; so I switched to auxiliary damping, which controlled the spacecraft until the automatic fuel was also expended. I was reaching for the switch to deploy the drogue parachute early in order to reduce these reentry oscillations, when it was deployed automatically. The drogue parachute stabilized the spacecraft rapidly.

At 10,800 feet the main parachute was deployed. I could see it stream out behind me, fill partially, and then as the reefing line cutters were actuated it filled completely. The opening of the parachute caused a jolt, but perhaps less than I had expected.

The landing deceleration was sharper than I had expected. Prior to impact I had disconnected all the extra leads to my suit, and was ready for rapid egress, but there was no need for this. I had a message that the destroyer Noa would pick me up within 20 minutes. I lay quietly in the spacecraft trying to keep as cool as possible. The temperature inside the spacecraft did not seem to diminish. This, combined with the high humidity of the air being drawn into the spacecraft kept me uncomfortably warm and perspiring heavily. Once the Noa was alongside the spacecraft, there was little delay in starting the hoisting operation. The spacecraft was pulled part way out of the water to let the water drain from the landing bag.

During the spacecraft pickup, I received one good bump. It was probably the most solid jolt of the whole trip as the spacecraft swung against the side of the ship. Shortly afterwards the spacecraft was on the deck.

I had initially planned egress out through the top, but by this time I had been perspiring heavily for nearly 45 minutes. I decided to come out the side hatch instead.

General Remarks

Many things were learned from the flight of Friendship 7. Concerning spacecraft systems alone, you have heard many reports today that have verified previous design concepts or have shown weak spots that need remedial action. Now, what can be said of man in the system?

Reliability

Of major significance is the probability that much more dependence can be placed on the man as a reliably operating portion of the man spacecraft combination. In many areas his safe return can be made dependent on his own intelligent actions. Although a design philosophy could not be followed up to this time, Project Mercury never considered the astronaut as merely a passive passenger.

These areas must be assessed carefully, for man is not infallible, as we are all acutely aware. As an in-flight example, some of you may have noticed a slight discrepancy between launch photographs of the pilot and similar reentry views. The face plate on the helmet was open during the reentry phase. Had cabin pressure started to drop, I could have closed the face plate in sufficient time to prevent decompression, but nevertheless a face-plate open reentry was not planned.

On the ground, some things would also be done differently. As an example, I feel it more advisable in the event of suspected malfunctions, such as the heat-shield-retropack difficulties, that require extensive discussion among ground personnel to keep the pilot updated on each bit of information rather than waiting for a final clear cut recommendation from the ground. This keeps the pilot fully informed if there would happen to be any communication difficulty and it became necessary for him to make all decisions from onboard information.

Many things would be done differently if this flight could be flown over again, but we learn from our mistakes. I never flew a test flight on an airplane that I didn't return wishing I had done some things differently. Even where automatic systems are still necessary, mission reliability is tremendously increased by having the man as a backup. The flight of Friendship 7 is a good example. This mission would almost certainly not have completed its three orbits, and might not have come back at all if a man had not been aboard.

Adaptability

The flight of the Friendship 7 Mercury spacecraft has proved that man can adapt very rapidly to this new environment. His senses and capabilities are little changed in space. At least for the 4.5-hour duration of this mission, weightlessness was no problem. Man's adaptability is most evident in his powers of observation. He can accomplish many more and varied experiments per mission than can be obtained from an unmanned vehicle. When the unexpected arises as happened with the luminous particles and layer observations on this flight, he can make observations that will permit more rapid evaluation of these phenomena on future flights. Indeed, on an unmanned flight there likely would have been no such observations.

Future Plans

Most important, however, the future will not always find us as power limited as we are now. We will progress to the point where missions will not be totally preplanned. There will be choices of action in space, and man's intelligence and decision-making capability will be mandatory. Our recent space efforts can be likened to the first flights at Kitty Hawk. They were first unmanned but were followed by manned flights, completely preplanned and of a few seconds duration. Their experiments were, again, power limited, but they soon progressed beyond that point. Space exploration is now at the same stage of development. From all of the papers in this volume, I am sure you will agree with me that some big steps have been taken toward accomplishing the mission objectives expressed at the beginning of this paper.

13. SUMMARY OF RESULTS

By GEORGE M. LOW, Director of Space Craft and Flight Missions, Office of Manned Space Flight, National Aeronautics and Space Administration.

The fact that John Glenn's flight was an unqualified success is well documented in the preceding papers. This flight marked a major milestone in the United States program for the manned exploration of space. It would seem to be appropriate, at this time, to sum up what has been learned during this first phase and to interpret the results in terms of future missions.

In the fall of 1958 the stated objective of Project Mercury was to: "Determine man's capabilities in a space environment." This objective has been achieved for the missions accomplished to date. Man's reactions to the accelerations of launch and to the decelerations of reentry have been learned. It has been determined that a trained pilot can perform tasks under a relatively high g-stress as well as under zero-g, can monitor all his systems, can manually control the flight sequence, and can adequately control the attitude of his craft. The period of weightlessness has been extended by more than two orders of magnitude from 1 minute to nearly 300 minutes. It has been learned that there are no deleterious psychological or physiological effects resulting from this prolonged exposure to weightlessness, even though attempts were made to induce such effects.

In Project Mercury, far more has been learned than was anticipated — far more than merely the determination of man's capabilities in space. A knowledge of how to design, develop and manufacture a craft specifically engineered for man's flight into space has been gained. It has been learned how, through an intensive ground and flight test program, such a spacecraft can be developed to carry out its assigned mission. Ways have been determined to modify existing launch vehicles, designed for other purposes, to make them suitable for manned flight. The development of an abort sensing system, together with the most stringent quality control, has permitted the use of the Atlas missile in a program for which it was not designed or developed.

A knowledge of how to implement an extensive network of tracking stations, a network which is unique in that it makes use of real-time data transmission and real-time computing, and thereby permits real-time flight control, has been gained. Ground rules have been established for recovery from space. It has been learned how ships and aircraft, with information provided by the tracking network, can locate and retrieve a spacecraft after it has landed.

Some of the items that were developed for Project Mercury will find use in other fields. For example, the new lightweight survival equipment might well be used by Air-Rescue services throughout the world. The biomedical instrumentation for measuring respiration rate, temperature, activity of the heart, and blood pressure and for transmitting these quantities over long distances may also find uses in fields other than the exploration of space.

Extensive training and simulation has been found to be an absolute requirement. The training of the pilots has, of course, received a great deal of attention. Equally important is the extensive simulation of flights carried out by all persons involved in an actual operation. All the flight controllers and the network, computer, and communications experts have performed literally hundreds of practice missions wherein every conceivable emergency was simulated. Through these exercises, they have learned to work together as a well-functioning team, a team that supports the pilot throughout his mission.

Most important of all, it has been learned that a well-trained pilot, like Shepard, Grissom, or Glenn or like the other astronauts, can perform a mission in space just as well as he can perform a mission in the earth's atmosphere. The knowledge derived in the last 3 years is tremendous. Yet, in recognizing this fact, it must also be recognized that manned space flight is still in its earliest development stages. The flights of Shepard, Grissom, and Glenn were pioneering ventures and, as such, were not undertaken without risk. In accepting the challenge of future flights, in Projects Mercury, Gemini, and Apollo, it should not be forgotten that the risk in these missions will be at least as great as it has been in the past.

APPENDIX A
MERCURY NETWORK PERFORMANCE SUMMARY FOR MA-6

By THE MANNED SPACE FLIGHT SUPPORT DIVISION, NASA Goddard Space Flight Center

Summary

The performance of the Mercury Network was considered highly successful for the Mercury/Atlas-6 mission. At the time of launch, 14:47:39Z on February 20, 1962, all systems required to support the flight were operational. This was phenomenal considering the vast amount of equipment committed to support the mission.

Radar

The Mercury Network includes both C-band and S-band radars located around the Mercury ground track in such a manner that redundancy is afforded in case of a spacecraft beacon failure. The radars have a range capability of approximately 500 miles for the C-band radars and 1,000 miles for the S-band radars. During this mission, all radar sites tracked the spacecraft with C-band and/or S-band radars when it was within range. Data were supplied, in real time, to the dual Goddard computers at the rate of one data point per 6-second interval. An average of about 50 radar data points was received from each site with as many as 93 points from several. A majority of the sites tracked the spacecraft from one horizon to the other. The tracking was of such quality that the Goddard computers were supplied with more than enough data to update the orbital parameters for each orbit. The quality of the network data is indicated by the following typical values of standard error (eliminating all data points for pointing angles below 3° elevation).

Woomera FPS - 16
Data points	85
Range, yd	6.9
Azimuth, mil	0.08
Elevation, mil	0.25

Bermuda FPS - 16
Data points	50
Range, yd	8.6
Azimuth, mil	0.17
Elevation, mil	0.49

Muchea Verlort
Data points	93
Range, yd	17.6
Azimuth, mil	0.97
Elevation, mil	0.81

California Verlort
Data points	49
Range, yd	7.0
Azimuth, mil	0.60
Elevation, mil	0.90

A summary of the radar data received at Goddard during this mission (including all points) is shown in table A-1 (see ref. 1, table 12, p, 68) and the radar coverage times are shown graphically in figures A-1 to A-6.

Computing

Throughout the mission the automatic computing system at Goddard effectively used the network data, to supply real-time digital display and plot board information to the Mercury Control Center at Cape Canaveral. During launch the high-speed data from the Cape to Goddard were uninterrupted and of good quality, and the flight parameters were such that the computer recommended a GO. The Goddard computers quickly established the orbit from early network data and supplied real-time acquisition data to all sites. The precision of the data was indicated by the fact that, the time of retrofire, as recommended by the computers, was adjusted by only 2 seconds during the entire mission. During the reentry phase of the mission, the network data permitted a computation of predicted landing point which varied by only 2 miles.

Station	Radar	Total possible Valid observations	Valid observations	Nonvalid observations	Range, yd	Azimuth, mils	Elevation, mils
TABLE A-I. — Orbital Data Analysis, Radar Tracking							
					Standard Deviation		
First pass:							
BDA	FPS-16	71	52	3	34.5	0.34	0.61
BDA	Verlort	71	47	8	26.8	1.7	1.6
CYI	Verlort	68	63	9	73.3	1.4	2.0
MUC	Verlort	82	76	6	23.4	1.94	1.47
WOM	FPS-16	40	39	1	6.9	.077	.25
GYM	Verlort	65	51	14	30.3	1.02	1.82
WHS	FPS-16	34	28	6	5.9	.22	1.20
TEX	Verlort	64	46	18	71.0	2.69	1.60
EGL	FPS-16	40	38	1	8.58	.240	.477
EGL	MPQ-31	65	18	47	38.7	2.85	2.35
Second pass:							
CNV	FPS-16	64	43	21	39.3	.288	.871
BDA	FPS-16	66	51	15	14.5	.380	.656
BOA	Verlort	DATA NOT AVAILABLE					
CYI	Verlort	54	48	6	27.1	1.71	1.54
MUC	Verlort	80	60	20	39.1	1.43	1.32
WOM	FPS-16	33	28	5	2.29	.081	.13
HAW	FPS-16	15	15	0	4.92	.313	.251
HAW	Verlort	56	45	11	80.1	1.89	1.35
CAL	FPS-16	38	28	10	8.88	.544	.208
	Verlort	47	27	20	20.1	.705	.800
WHS	FPS-16	41	31	10	17.2	.179	.669
TEX	Verlort	60	58	2	41.5	1.61	1.41
EGL	FPS-16	40	32	8	6.85	.209	.286
	MPQ-31	50	32	18	103.8	3.82	3.74
Third pass:							
CNV	FPS-16	64	26	38	57.3	.318	.936
BDA	FPS-16	65	56	9	30.6	.157	.561
	Verlort	DATA NOT AVAILABLE					
CYI	Verlort	OUT OF RANGE					
MUC	Verlort-	70	69	1	31.6	.803	1.24
WOM	FPS-16	OUT OF RANGE					
HAW	FPS-16	38	37	1	8.05	.230	.557
	Verlort	64	52	12	38.6	2.07	1.60

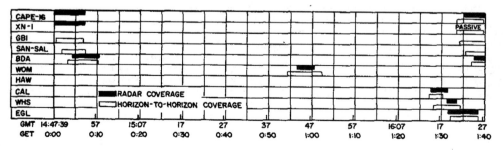

FIGURE A-1. C-band radar coverage, first orbit.

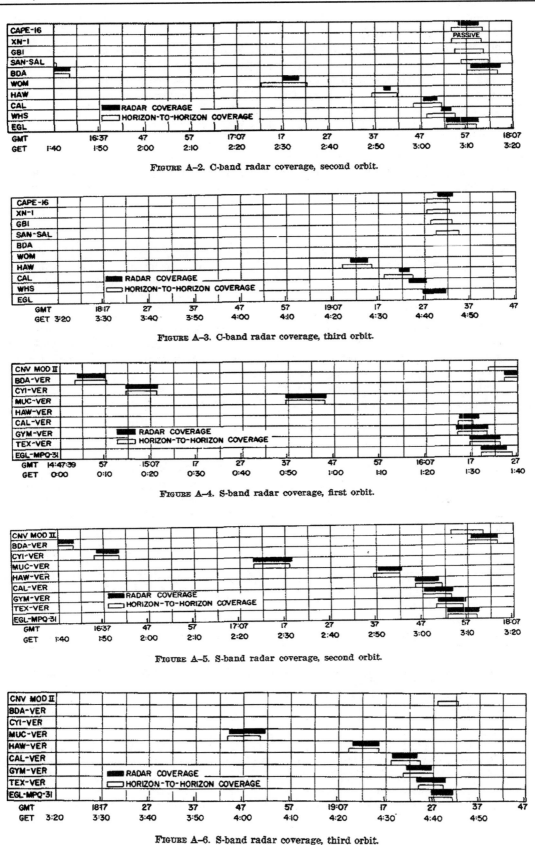

FIGURE A-2. C-band radar coverage, second orbit.

FIGURE A-3. C-band radar coverage, third orbit.

FIGURE A-4. S-band radar coverage, first orbit.

FIGURE A-5. S-band radar coverage, second orbit.

FIGURE A-6. S-band radar coverage, third orbit.

Acquisition Aid

The automatic acquisition aid subsystems performed as expected with no major problems encountered. As usual, multipath was a problem at low elevation angles and, therefore manual elevation control was used. Four sites, Canary Island, Muchea, California, and Texas, used real-time computed pointing data for direct radar acquisition, independent of the automatic acquisition systems. This was excellent verification of the accuracy of acquisition data furnished to the network radars by the Goddard computers.

Command

The command subsystems operated in a satisfactory manner for the mission with a total of eleven functions being successfully transmitted to the spacecraft from various sites.

Telemetry

The telemetry subsystem reception and performance was outstandingly good. All stations acquired and lost signals at or near the horizon. No major operator error or equipment malfunction was reported that influenced mission monitoring and control. The maximum range of telemetry reception varied from 500 to 1,100 miles.

The malfunction of the landing-bag-deploy microswitch was first indicated by the telemetry system as the spacecraft passed Cape Canaveral at the end of the first orbit. Since this event is normally not displayed, remote sites were requested to monitor this function on the events recorders for the remainder of the mission. After a number of sites confirmed that this event was indicated, the astronaut was informed and given a course of action. A summary of telemetry subsystem performance is shown in tables A-II to A-VII and the telemetry coverage times are shown graphically in figures A-7 to A-9.

Voice Communication

Voice communication between the ground and the spacecraft was considered excellent. The quality of the air-ground voice communication monitored on the Goddard conference loop was very good and provided the flight controllers at Mercury Control Center with adequate monitoring capability throughout the mission. The coverage times of HF and UHF communications are shown graphically in figures A-10 to A-12.

Timing

The timing system performed very well with the, exception of the serial decimal GMT time used on the strip-chart recorders at Hawaii and Kano. The real-time records from these sites are usable in spite of the timing malfunctions.

Data Transmission

No problems were encountered with the data transmission system; all high-speed and low speed data lines were operational during the entire mission.

Ground Communication

The teletype system circuits performance was good with relatively few outage periods. Traffic flow was exceptionally smooth, with transmission times generally less than 1 minute. There were 2,048 lines of radar data automatically transmitted by teletype with only 15 lines in error. All acquisition messages before retro-fire were dispatched to sites in time for effective use.

TABLE A–II.—*Telemetry Data, Orbit 1*

Station	Telemetry		Decommutator		Slant range, naut. miles		Elevation, deg	
	Acquisition of signal	Loss of signal	Lock	Loss	Acquisition of signal	Loss of signal	Acquisition of signal	Loss of signal
BDA	00:03:02	00:10:26	00:03:40	00:10:26	750	868	0	−1.2
ATS			Not applicable					
CYI	00:14:15	00:21:23	00:14:41	00:21:20	800	850	0	0
KNO	00:21:13	00:28:21	00:21:50	00:28:21	850	900	−.3	−.5
ZZB	00:29:51	00:37:51	00:30:01	00:38:01	920	990	−.2	−.6
IOS	00:40:02	00:48:31	00:43:12	00:46:56	1000	1040	−.6	−1
MUC	00:49:21	00:57:55	00:49:32	00:57:21	1020	990	−.4	−8
WOM	00:54:00	01:02:41	00:54:16	01:02:37	810	1060	+3	−1.5
CTN	01:09:19	01:17:42	01:09:36	01:17:40	900	1150	+.3	−4
HAW			Not applicable					
CAL	01:26:41	01:31:23	01:27:18	01:31:23	840	920	−.7	−2.1
GYM	01:26:47	01:33:25	01:27:01	01:33:15	730	950	+.9	−2.5
WHS			Not applicable					
TEX	01:29:24	01:36:18	01:29:32	01:36:14	830	820	−.7	−.6
EGL	01:32:00	01:37:05	01:32:11	01:37:00	800	880	−1	−1.5

TABLE A–III.—*Telemetry Data, Orbit 2*

Station	Telemetry		Decommutator		Slant range, naut. miles		Elevation, deg	
	Acquisition of signal	Loss of signal	Lock	Loss	Acquisition of signal	Loss of signal	Acquisition of signal	Loss of signal
BDA	01:36:38	01:43:53	01:36:49	01:43:53	860	890	−1.2	−1.4
ATS	01:51:54	01:58:31	01:53:04	01:58:21	880	830	−.2	+1
CYI	01:47:55	01:53:58	01:48:11	01:53:53	850	910	−.2	−.2
KNO	01:54:47	02:01:21	01:55:07	02:01:21	890	940	−.6	−.6
ZZB	02:04:05	02:10:51	02:04:13	02:10:51	920	1040	.23	−1.1
IOS	02:12:17	02:22:09	02:13:27	02:21:54	1100	1050	−1.9	−.9
MUC	02:22:51	02:31:23	02:23:06	02:31:22	1008	960	−.3	0
WOM	02:27:36	02:35:45	02:27:45	02:35:39	950	1020	+.5	−1.5
CTN	02:42:51	02:49:45	02:42:59	02:49:38	870	907	−1.5	−1.3
HAW	02:49:01	02:55:19	02:49:29	02:55:08	940	830	−1.3	−.8
CAL	02:58:11	03:04:48	02:58:35	03:04:48	880	730	−1.5	+.7
GYM	02:59:59	03:06:44	03:00:13	03:06:34	610	880	+3	−1.5
WHS			Not applicable					
TEX	03:03:14	03:09:39	03:03:16	03:09:31	810	810	−.8	−.4
EGL	03:05:35	03:12:07	03:05:46	03:12:00	670	1000	+2	−3

TABLE A–IV.—*Telemetry Data, Orbit 3*

| Station | Telemetry | | Decommutator | | Slant range, naut. miles | | Elevation, deg | |
	Acquisition of signal	Loss of signal	Lock	Loss	Acquisition of signal	Loss of signal	Acquisition of signal	Loss of signal
BDA	03:09:56	03:17:03	03:10:06	03:17:03	870	900	−1.2	−1.2
ATS	03:24:44	03:32:25	03:25:06	03:31:22	900	920	−.5	+.5
CYI								
KNO			Not applicable					
ZZB								
IOS	03:46:55	03:56:49	03:48:10	03:56:30	1050	1100	−1	−1.4
MUC	03:56:31	04:04:12	03:56:49	04:04:08	1020	940	−.7	−.16
WOM	04:03:16	04:06:19	04:03:31	04:06:01	870	1000	+1	1.1
CTN			Not applicable					
HAW	04:21:49	04:28:49	04:22:02	04:24:39	922	770	−2	+.1
CAL	04:31:17	04:37:57	04:31:27	04:37:56	900	540	−2	+3.6
GYM	04:33:44	04:39:49	04:34:04	04:39:39	770	740	−5	−1.1
WHS			Not applicable					
TEX	04:36:53	04:42:32	04:36:58	04:42:34	930	603	−3	−5
EGL	04:39:00	04:42:52	04:39:21	04:42:48	800	500	−1	+1.4

TABLE A–V.—*Telemetry Receiver Signal Strength, Orbit 1*

| Station | Estimated mean, microvolts | | | |
	Low (Receiver 1, model 1415)	Low (Receiver 2, model 1434)	High (Receiver 1, model 1415)	High (Receiver 2, model 1434)
MCC	Not applicable			
BDA	250	250	250	250
CYI	75	120	25	190
ATS	No contact			
KNO	100	50	100	80
ZZB	110	92	84	134
IOS	20	80	35	150
MUC	205	205	259	259
WOM	200	200	210	210
CTN	30	70	25	25
HAW	No contact			
GYM	130	150	150	250
CAL	40	20	40	10
WHS	No telemetry equipment			
TEX	100	180	90	200
EGL	No telemetry equipment			

TABLE A–VI.—*Telemetry Receiver Signal Strength, Orbit 2*

| Station | Estimated mean, microvolts | | | |
	Low (Receiver 1, model 1415)	Low Receiver 2, model 1434)	High (Receiver 1, model 1415)	High (Receiver 2, model 1434)
MCC	Not applicable			
BDA	250	250	250	250
CYI	40	180	50	80
ATS	015	15	10	60
KNO	70	60	70	50
ZZB	71	32	43	60
IOS	40	150	80	150
MUC	204	204	204	204
WOM	100	130	200	100
CTN	40	60	30	35
HAW	90	50	60	80
GYM	120	200	100	160
CAL	80	50	80	30
WHS	No telemetry equipment			
TEX	90	100	70	200
EGL	No telemetry equipment			

TABLE A-VII.—*Telemetry Receiver Signal Strength, Orbit 3*

Station	Estimated mean, microvolts			
	Low (Receiver 1, model 1415)	Low (Receiver 2, model 1434)	High (Receiver 1, model 1415)	High (Receiver 2, model 1434)
MCC	Not applicable			
BDA	200	200	200	200
CYI	No contact			
ATS	40	30	10	200
KNO	No contact			
ZZB				
IOS	40	125	60	125
MUC	64	104	54	133
WOM	35	32	68	30
CTN	No contact			
HAW	222	225	200	200
GYM	90	80	80	80
CAL	80	40	80	40
WHS	No telemetry equipment			
TEX	40	225	90	200
EGL	No telemetry equipment			

Very good support by the voice network permitted exceptionally fine communications between the Mercury Control Center and sites with voice terminations. Echo was reported on the Guaymas line at the T-33 voice check; a speaker was found to be feeding back into the system. Appropriate action was taken to correct this condition promptly.

Conclusion

It is concluded that there were no major network problems encountered during the MA-6 mission. However, there were a number of minor problems, as indicated, which are currently under investigation.

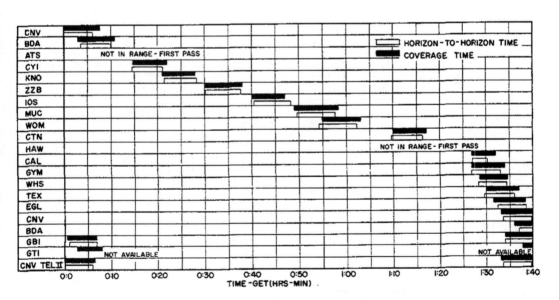

FIGURE A-7. Telemetry reception coverage, first orbit.

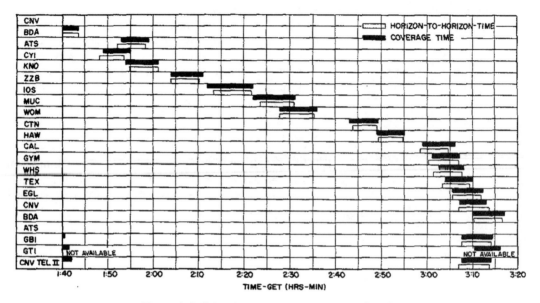

FIGURE A–8. Telemetry reception coverage, second orbit.

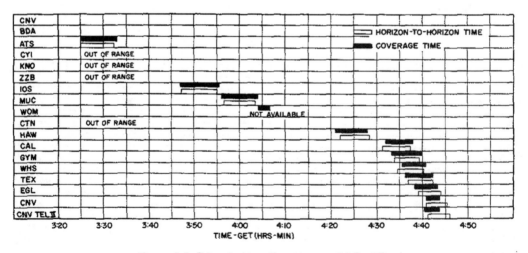

FIGURE A–9. Telemetry reception coverage, third orbit.

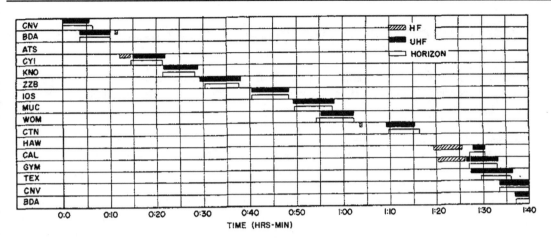

FIGURE A-10. HF and UHF communications, first orbit.

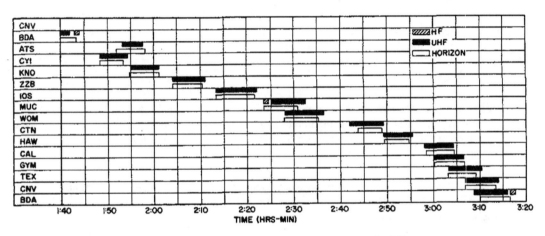

FIGURE A-11. HF and UHF communications, second orbit.

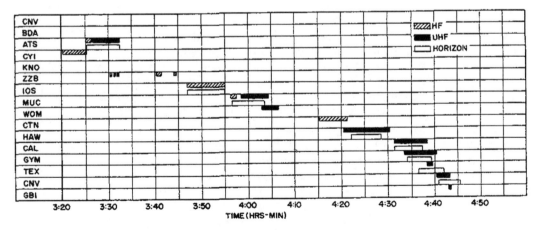

FIGURE A-12. HF and UHF communications, third orbit.

APPENDIX B —
AIR-GROUND COMMUNICATIONS OF THE MA-6 FLIGHT

The following table is a verbatim transcription of the MA-6 flight communications taken from the spacecraft onboard tape recording. This is therefore a complete transcript of the communications received and transmitted by Astronaut Glenn.

In a few cases, the communications do not agree with post flight detailed analysis of telemetered and recorded data. No attempt has been made to correct the transcript, and, therefore, the technical papers should be considered authoritative in the event of conflict.

In the table, column one is the elapsed time from the launch of the spacecraft in hours, minutes, and seconds that the communiqué was initiated. Column two is the duration in seconds of the communiqué. Column three identifies the communicator as follows:

CC— Capsule (spacecraft) Communicator at the range station
P — Pilot (astronaut)
CT— Communications Technician at the range station
S — Surgeon or medical monitor at the range station
SY— Systems monitor at the range station
R — Recovery personnel

All temperatures are given as °F; all pressures are in pounds per square inch absolute (psia) ; fuel, oxygen, and coolant quantities are expressed in remaining percent of total nominal capacities; retrosequence times are expressed in hours, minutes, and seconds (i.e., 04 52 47 means 4: hours, 52 minutes, and 47 seconds from instant of lift-off).

Within the text, a series of dots were used to designate times where communiqués could not be deciphered. The station in prime contact with the astronaut and the orbit number are designated at the initiation of communications with that station.

CAPE CANAVERAL (FIRST ORBIT)

CC			3,2,1,0
00 00 03	4.0	P	Roger. The clock is operating. We're underway.
00 00 07	1.5	CC	Hear loud and clear.
00 00 08	2.0	P	Roger. We're programming in roll okay.
00 00 13	3.5	P	Little bumpy along about here.
00 00 15	1.0	CC	Roger.
00 00 17	2.0	CC	Standby for 20 seconds.
00 00 19	0.5	P	Roger.
00 00 20	2.0	CC	2 - 1, mark.
00 00 23	3.5	P	Roger. Backup clock is started.
00 00 32	5.0	P	Fuel 102-101 [percent], oxygen 78-100 [percent], amps 27.
00 00 39	3.0	CC	Roger. Loud and clear. Flight path is good, 69 [degrees].
00 00 43	4.0	P	Roger. Checks okay. Mine was 70 [degrees] on your mark.
00 00 48	2.5	P	Have some vibration area coming up here now.
00 00 52	2.0	CC	Roger. Reading you loud and clear.
00 00 55	4.0	P	Roger. Coming into high Q a little bit; and a little contrail went by the window or something there.
00 01 00	0.5	CC	Roger.
00 01 03	6.5	P	Fuel 102-101 [percent], oxygen 78-101 [percent], amps 24. Still Okay.
00 01 12	3.0	P	We're smoothing out some now, getting out of the vibration area.
00 01 16	3.0	CC	Roger. You're through max. Q. Your flight path is . . . .

00 01 19	3.2	P	Roger. Feels good, through max. Q and smoothing out real fine.
00 01 26	4.0	P	Cabin pressure coming down by 7.0 okay; flight very smooth now.
00 01 31	2.0	P	Sky looking very dark outside.
00 01 42	3.0	P	Cabin pressure is holding at 6.1 okay.
00 01 46	3.5	CC	Roger. Cabin pressure holding at 6.1.
00 01 49	3.5	P	Roger. Have had some oscillations, but they seem to be damping out okay now.
00 01 56	9.5	P	Coming up on two minutes, and fuel is 102-101 [percent], oxygen 78-102 [percent]. The g's are building to 6.
00 02 07	5.0	CC	Roger. Reading you loud and clear. Flight path looked good. Pitch 25 [degrees] Standby for.
00 02 12	8.0	P	Roger. BECO, back to 1¼ g's. The tower fired; could not see the tower go. I saw the smoke go by the window.
00 02 21	2.0	CC	Roger. We confirm staging on TM.
00 02 24	0.5	P	Roger.
00 02 27	3.0	P	Still have about 1½ g's. Programming. Over.
00 02 36	7.5	P	There the tower went right then. Have the tower in sight way out. Could see the tower go. Jettison tower is green.
00 02 48	0.3	CC	Roger.
00 02 50	0.5	P	1½ g's.
00 02 63	3.5	CC	Roger, Seven. Still reading you loud and clear. Flight path looks good.
00 02 56	6.0	P	Roger. Auto Retro jettison is off; Emergency Retro jettison Fuse switch, off; Retro jettison Fuse switch, off.
00 03 03	1.5	P	UHF/DF to normal.
00 03 19	2.3	CC	Flight path looks good; steering is good.
00 03 22	5.0	P	Roger. Understand everything looks good; g's starting to build again a little bit.
00 03 30	0.5	CC	Roger.
00 03 32	1.5	CC	Friendship Seven. Bermuda has you.
00 03 34	13.0	P	Roger. Bermuda standby. This is Friendship Seven. Fuel 103-101 [percent], oxygen 78-100 [percent]. All voltages above 25, amps 26.
00 03 48	4.5	CC	Roger. Still reading you loud and clear. Flight path is very good. Pitch, -3 [degrees].
00 03 53	0.5	P	Roger.
00 03 56	3.0	P	My pitch checks a - 7 [degrees] on your - 3 [degrees].
00 04 00	0.8	CC	Roger, Seven.
00 04 08	10.5	P	Friendship Seven. Fuel 103-101 [percent], oxygen 78-100 [percent], amps 25, cabin pressure holding at 5.8.
00 04 20	5.0	CC	Roger. Reading you loud and clear. Seven, Cape is Go; we're standing by for you.
00 04 25	16.5	P	Roger. Cape is Go and I am Go. Capsule is in good shape. Fuel 103-102 [percent], oxygen 78-100 [percent], cabin pressure holding steady at 5.8, amps is 26. All systems are Go.
00 04 44	2.0	CC	Roger. 20 seconds to SECO.
00 04 47	0.5	P	Roger.
00 04 49	1.5	P	Indicating 6 g's.
00 04 52	0.5	P	Say again.
00 04 53	1.0	CC	Still looks good.
00 04 54	0.5	P	Roger.
00 05 04	4.0	P	SECO, posigrades fired okay.
00 05 10	0.5	CC	Roger, stand . . . .
00 05 12	5.0	P	Roger. Zero-g and I feel fine. Capsule is turning around.
00 05 18	1.8	P	Oh, that view is tremendous!
00 05 21	1.5	CC	Roger. Turnaround has started.
00 05 23	7.0	P	Roger. The capsule is turning around and I can see the booster during

			turnaround just a couple of hundred yards behind me. It was beautiful.
00 05 30	4.5	CC	Roger, Seven. You have a go, at least 7 orbits.
00 05 35	4.5	P	Roger. Understand Go for at least 7 orbits.
00 05 44	7.0	P	This is Friendship Seven. Can see clear back; a big cloud pattern way back across towards the Cape. Beautiful sight.
00 05 54	3.5	CC	Roger, still reading you loud and clear. Next transmission, Bermuda.
00 05 58	10.5	P	Roger. Understand next transmission, Bermuda. Capsule did damp okay and turned around. Scope has extended, okay. Taking off the filter I had on it before launch.
00 06 02	1.0	P	Making electrical check.
00 06 18	4.5	P	All batteries 25 or above, on main. Going through orbit checklist.

BERMUDA (FIRST ORBIT)

00 06 25	2.0	CC	Roger, Friendship Seven. Orbit checklist.
00 06 27	6.0	P	Landing Bag is off. Emergency Retrosequence, off. Emergency Drogue Deploy is off.
00 06 38	1.5	CC	Emergency Landing Bag is next.
00 06 40	17.0	P	Roger. Landing Bag was already off. I got it first and reported it. Retromanual is off, and we're all set. This is very comfortable at Zero-g. I have nothing but very fine feeling. It just feels very normal and very good.
00 06 58	2.0	CC	Friendship Seven. Standby for retrosequence times.
00 07 00	1.0	P	Roger. Ready to copy.
00 07 08	4.2	CC	Roger. 1B, 00 17 50.
00 07 08	4.3	P	Roger. 1 Bravo is 00 plus 17 plus 50.
00 07 14	3.6	CC	End of orbit, 01 28 54.
00 07 19	2.9	P	Roger, 01 plus 28 plus 54.
00 07 22	3.8	CC	Roger, end of mission is 04 32 47.
00 07 28	6.3	P	Roger. 04 plus 32 plus 47. Do I have okay for resetting clock? Over.
00 07 36	8.5	CC	Negative, don't reset the clock. Your V over Vr is unity, your apogee altitude is 138 [nautical miles] and are you starting your control systems check?
00 07 45	5.6	P	Roger. As soon as we get done with this transmission. I understand am I cleared to control systems check.
00 07 51	1.7	P	Roger. Starting controls check.
00 08 01	1.1	P	Starting controls check.
00 08 09	1.3	CC	Your attitudes look okay here.
00 08 32	2.4	CC	Friendship Seven. Anything to report on control systems checks.
00 08 34	6.9	P	Not yet, everything appears to be going okay. Am now on the yaw part of the check. Going off right on schedule.
00 08 42	0.8	CC	Very good, very good.
00 08 50	4.5	P	Control so far is excellent. Very good, no problems at all so far on control.
00 08 55	0.6	CC	Roger.
00 09 01	2.7	P	Aux Damp pulls it right in every time. No problems.
00 09 04	0.4	CC	Very good.
00 09 24	1.8	CC	Friendship Seven, Bermuda. Do you still read?
00 09 26	5.6	P	Roger, Bermuda, still read you loud and clear. Still completing control check. Having no problem at all so far.
00 09 32	1.4	CC	Roger, you're still loud and clear.
00 09 35	0.5	P	Roger.
00 09 46	10.8	P	This is Friendship Seven. Working just like clockwork on the control check, and it went through just about like the Procedures Trainer runs. It's very smooth and I checked ....
00 09 58	1.9	CC	On UHF, if you read, go to HF.

00 10 00	1.1	P	Roger. Going.
00 10 24	2.8	CC	Friendship Seven, Bermuda CapCom on HF.
00 10 27	3.5	P	Hello, Hello, Bermuda. Receive you loud and clear; how me?
00 10 38	2.1	CC	Friendship Seven, Friendship Seven, Bermuda CapCom on HF.
00 10 40	6.7	P	Hello, Bermuda, Hello, Bermuda CapCom. Friendship Seven reads you loud and clear. Hello, Bermuda, Friendship Seven. How me?
00 10 53	2.2	CC	Friendship Seven, this Bermuda CapCom on HF.
00 10 56	4.4	P	Hello, Bermuda CapCom, this is Friendship Seven. Read you loud and clear; how me?

CANARY (FIRST ORBIT)

00 11 16	1.9	P	Hello, Canary, Friendship Seven. Over.
00 11 20	5.3	CC	Friendship Seven, Friendship Seven, this is Canary CapCom. Read you loud and clear. Over.
00 11 26	14.0	p	Canary, Friendship Seven. Roger. Control check complete. Capsule in ASCS and holding. I have booster in sight out the window. It's probably about 1 mile away and going down under my position and a little bit to my left. Over.
00 11 46	7.3	P	This is Friendship Seven. Everything is still Go. Capsule is in fine shape. Holding pressure at 5.8. Over.
00 12 18	3.4	CC	Friendship Seven, Friendship Seven, this is Canary Cap Com. Do you read? Over
.00 12 21	3.5	P	Hello, Canary Cap Com, loud and clear on HF. How me? Over.
00 12 26	7.1	CC	I read you loud and clear. Would you give me your fuel on your control systems check.
00 12 34	6.7	P	Roger. This is Friendship Seven. Control systems check was perfect. Control systems checks perfect. Over.
00 12 42	3.4	CC	Roger, Friendship Seven, understand control systems check was perfect.
00 12 47	1.1	P	That's affirmative.
00 12 52	5.2	CC	Friendship Seven, this Canary Cap Com. Could you get started with your station report? Over.
00 13 00	2.4	P	Hello, Canary, Friend Seven. Repeat please. Over.
00 13 05	4.3	CC	What is your space, spacecraft station, status report?
00 13 10	3.9	P	This is Friendship Seven. Standby. I'm getting out some equipment. Over.
00 14 03	3.7	P	Hello, Canary, Friendship Seven, switching to UHF. Over.
00 14 09	4.6	CC	Friendship Seven, Friendship Seven, Canary Cap Com. Say again.
00 14 23	4.3	CC	Friendship Seven, Friendship. Seven, Canary Cap Com. Read you.
00 14 32	3.5	P	Hello, Canary, Friendship Seven on UHF. How now? Over.
00 14 37	2.1	CC	I read you loud and clear.
00 14 39	5.9	P	Roger, understand loud and clear. I read you much better also, than I did on HF. Over.
00 14 46	1.3	CC	Same here.
00 14 48	3.6	P	This is Friendship Seven. Everything is going fine.
00 14 55	3.2	P	Getting, getting some of the equipment together. Over.
00 14 59	0.5	CC	Roger.
00 15 12	5.2	CC	Friendship Seven, Friendship Seven. What is your spacecraft status report? Over.
00 15 17	1:0.4	P	Roger. Standby, Will give status report. I am in orbit attitude for your tracking. Status report follows: Fuses all number one except Tower Sep number two. Emergency Retrosequence, Emergency Retrojettison, and Emergency Drogue are in the center-off position. Squib is Armed. Auto Retrojettison is off. ASCS is normal, auto, gyro normal. All "T" handles are in the "in"-position. Retro Delay is normal. Cabin Lights are on both. Photo Lights are still on. Telemetry Low Frequency is on. Rescue

			Aids are on automatic. Ah, Jettison Tower, and Sep Capsule lights are out. The pressure regulator is still in the "in" position. Launch control is on. All sequence panel positions are normal except Landing Bag is off. Are you receiving? Over.
00 16 19	1.8	CC	Understand. We have telemetry solid.
	2.0	P	Roger, you have telemetry solid.
00 16 25	6.7	P	Control fuel is 90-98 [percent], I repeat, 90-98 [percent].
00 36 32	2.9	CC	Automatic fuel is 90 [percent], 98 [percent].
00 16 35	1:21.7	P	That is affirmative. Attitude: roll 0 [degrees], yaw 2 [degrees] right, pitch -33[degrees]. Rates are all indicating zero. I am on ASCS at present time. The clock is still set for time to ret, for retrograde time of 04 plus 32 plus 28. I have retrograde times okay from Bermuda. Cabin pressure holding steady at 5.7. Cabin air 90 [degrees]. Relative humidity, 30 [percent]. Coolant quantity is 68 [percent]. Suit environment is 65. Suit pressure is indicating 5.8. Steam temperature 60 [degrees] on the suit. I am very comfortable. However, I do not want to turn it down just yet. Primary oxygen is 78 [percent]; secondary 102 [percent]. Main bus is 24. Number one is 25, 25, 25. Standby one is 26; Standby two is 25; Isolated, 29, and back on main. Ammeter is indicating 23. ASCS is 112. Fans are 112. Over.
00 17 58	5.4	CC	I understand, I understand your retrosequence time is 04 32 28. Over.
00 18 04	20.7	P	That is affirmative. That's what's set in the clock. The horizon is a brilliant, a brilliant blue. There, I have the mainland in sight at present time coming up on the scope, and have Canaries in sight out through the window and picked them up on the scope just before I saw them out of the window. Over.
00 18 26	6.0	CC	Roger, Friendship Seven. This is Canary CapCom. Repeat blood pressure check, repeat blood pressure check.
0018 32	5.0	P	Roger, repeating blood pressure check now. Starting, pumping up.
0018 41	8.3	P	This is Friendship Seven. Have beautiful view of the African Coast, both in the scope, and out the window. Out the window is by far the best view.
00 18 56	2.7	P	Part of the Canaries was hidden by clouds
.00 19 09	2.8	CC	Roger, I read you. We're getting the blood pressure now.
00 19 12	1.8	P	Roger, Friendship Seven.
00 19 18	5.7	CC	Friendship Seven this is Canary CapCom. Your, your medical status is green; it looks okay.
00 19 24	3.2	P	Roger. This is Friendship Seven. What is blood pressure? Over.
00 19 30	0.4	CC	Standby.
00 19 33		CC	Friendship Seven. Your blood pressure is 120 over 80, repeat 120 over 80.
00 19 38	0.5	P	Roger.
00 19 40	8.2	CC	Friendship Seven, this is Canary CapCom. Your 150 volt-amp inverter is 175°, and holding.
00 19 49	1.6	P	Roger. Friendship Seven.
00 19 52	4.9	CC	Your 250 volt-amp inverter temperature is 150 [degrees], and holding.
00 20 02	1.0	P	Friendship Seven. Roger.
00 20 04	7.0	CC	Your auto fuel line temperature is 70°, your manual fuel line temperature is 100°.
00 20 12	1.9	P	Say again fuel temperatures. Over.
00 20 15	6.7	CC	Negative, your auto fuel line temperature was 70°, 70°.
00 20 22	1.1	P	Roger, very good.
00 20 25	1.9	CC	100°.
00 20 29	0.4	P	Roger.
00 20 33	8.8	P	This is Friendship Seven standing by. I am slightly behind on my checklist at present time. Will get caught up.
00 20 42	2.9	CC	Roger, this is Canary CapCom standing by.

| 00 21 16 | 11.4 | P | This is Friendship Seven still on ASCS. I can see dust storms down there blowing across the desert, a lot of dust; it's difficult to see the ground in some areas. Over. |
| 00 21 29 | 2.7 | CC | Seven, you are fading; how me? Over. |

KANO (FIRST ORBIT)

00 21 33	4.9	CC	Friendship Seven, this is Kano Cap Com, I read you loud and clear. How do you read me? Over.
00 21 37	2.2	P	Roger, Kano, loud and clear; how me?
00 21 40	3.3	CC	Roger, loud and clear. What is your status? Over.
00 21 44	3.5	P	This is Friendship Seven. My status is excellent. I feel fine. Over.
00 21 49	9.0	CC	Roger. I monitored part of your conversation over Canary and heard your comments on the weather over Africa. Will you give us status report? Over.
00 21 59	26.0	P	Roger, this is Friendship Seven. Fuel, 90-98 [percent]. Oxygen, 78-100 [percent]. Cabin pressure holding 5.6 at the present time. Have very little dirt floating around in the capsule, just a little bit and preferring to take xylose pill at present time. Unsealing the, going to unseal the faceplate. Over.
00 22 27	3.9	CC	Roger, Friendship Seven. Your exhaust temperature please. Over.
00 22 32	3.1	P	Say again. Standby one, taking xylose.
00 22 41	4.3	CC	Friendship Seven, this is Kano Cap Com. Can you give us the reading on your exhaust temperature?
00 22 49	1.7	P	Say again, Kano. Over.
00 22 51	2.3	CC	Reading on your exhaust temperature.
00 22 56	2.5	P	Roger. Steam temperature is 59 [degrees]. Over.
00 23 00	8.7	CC,	Roger. Friendship Seven. We have TM solid. If you want, we will standby while you do your yaw maneuver and check your systems.
00 23 11	3.5	P	Roger. Friendship Seven. I am taking xylose pill now.
00 23 19	1.9	CC	Roger, Friendship Seven, understand.
00 23 22	4.2	P	This is Friendship Seven, going to, UHF low for check. Over.
00 23 29	1.4	CC	Roger, Friendship Seven.
00 23 48	3.0	P	Hello, Kano. Friendship Seven. UHF Low; how now? Over.
00 23 53	2.6	CC	Friendship Seven, this is Kano Cap Com. Say again.
00 23 56	3.9	P	This is Friendship Seven on UHF Low. How do you receive me? Over.
00 24 00	3.0	CC	Friendship Seven, this is Kano. Read you loud and clear.
00 24 04	2.8	P	Roger, Kano, going back to UHF Hi. Over.
00 24 11	0.4	CC	Roger.
00 24 16	8.3	CC	We have a temperature of 189° on your 150 volt inverter and 150° your 250 volt inverter.
00 24 28	4.6	P	Roger. This is Friendship Seven on UHF Hi again. Understand inverter temperatures.
00 24 38	5.3	CC	Roger, Friendship Seven. We will standby while you do your yaw maneuver. Over.
00 24 43	9.5	P	Roger, this is Friendship Seven, starting in yaw maneuver. I'm about 40 seconds late on that one. Starting yaw maneuver at present time. Over.
00 24 54	1.1	CC	Roger, understand.
00 24 56	1.7	P	Going to manual control.
00 25 02	2.2	P	Correction, going to fly-by-wire. Over.
00 25 06	3.4	CC	Roger, understand on fly-by-wire for yaw maneuver.
00 25 09	0.7	P	That's affirmative.
00 25 30	16.5	P	This is Friendship Seven, having no trouble controlling on fly-by-wire. Drift is coming around at about 1° per second, and holding attitude okay in other axes. Over.

00 25 46	9.9	CC	Roger. Check drift at 1° per second and holding attitude okay all axes, I have your retrosequence time for Area I Charlie.
00 25 57	4.4	P	Standby. Will get it later; I'm in the middle of yaw maneuver at present time. Over.
00 26 02	0.6	CC	Standing by.
00 26 IS	5.7	P	This is Friendship Seven at 60° right yaw, and holding temporarily. Over.
00 26 25	2.5	CC	Roger. 60° right yaw and holding.
00 26 34	6.8	P	Attitudes all well within limits. I have no problem holding attitude with fly-by wire at all. Very easy. Over.
00 26 42	9.4	CC	Roger, check fly-by-wire is very easy. Our telemetry checks all your systems out okay. Are you ready for retrosequence time?
00 26 52	7.5	P	Negative, not yet. I'll pick it up possibly at next station if I lose contact with you. I'm still on manual control here on fly-by-wire. Over.
00 27 01	0.4	CC	Understand.
00 27 24	5.1	P	This is Friendship Seven, returning on fly-by-wire to orbit attitude. Over.
00 27 30	3.8	CC	Roger, Friendship Seven, I check your returning to orbit attitude.
00 27 34	8.6	P	Roger, this is Friendship Seven. Out the window, can see some fires down on the ground, long smoke trails right on the edge of the desert. Over.
00 27 44	6.1	CC	Roger. We've had dusty weather here, and as far as we can see, a lot of this part of Africa is covered with dust.
00 27 50	2.9	P	That's just exactly the way it looks from up here, too.
00 27 55	3.4	CC	Roger. You want to stand by for retrosequence time, Area I Charlie?
00 28 01	3.6	P	Roger. Going back on ASCS. Yaw check okay. Can pick up yaw fairly well in the scope. It's a little different display than I had really anticipated, but it checks okay, and I can pick up yaw. I have to be about 5° or so in yaw before I really start picking it up. Over.
00 28 26	11.3	CC	Seven, you are fading rapidly. I will broadcast this time in the blind for Area I Charlie, 00 32 12, 00 32 12.
00 28 40	4.2	P	00 32 22. Is that affirm? Over.

ZANZIBAR (FIRST ORBIT)

00 29 25	6.6	P	This is Friendship Seven. Hello, Zanzibar, Friendship Seven. Do you receive? Over.
00 29 39	6.2	CT	Friendship Seven, Friendship Seven, this is Zanzibar Com Tech transmitting on HF, UHF, do you copy? Over.
00 29 45	14.1	P	Roger, Zanzibar Com Tech, read you loud and clear. Control fuel is 90-98 [percent] cabin pressure, 5.6 and holding; oxygen, 75-100 [percent]. Over.
00 30 13	17.3	P	This is Friendship Seven in the blind for recording. Much of eastern Africa is covered by clouds, sort of wispy high cirrus looking clouds. Cannot see too much down there except the cloud decks themselves. Catch a sight of the ground underneath once in a while.
00 30 32	3.9	CC	Friendship Seven, this is Zanzibar Cap Com, reading you loud and clear.
00 30 37	1.2	P	Roger, Zanzibar.
00 30 39	7.2	CC	Message from IOS Cap Com, that he will not release balloon flare this orbit. Will fire parachute flares instead. Did you copy? Over.
00 30 47	2.1	P	Roger, this is Friendship Seven, understand.
00 30 52	10.1	CC	Friendship Seven, this is Zanzibar Cap Com. We have solid telemetry contact, report your status. Over.
00 30 58		P	Roger, this is Friendship Seven. Fuel 90-98 [percent], cabin pressure 55 and holding, oxygen 75-100 [percent], amps 24. Over.
00 31 12	5.3	CC	Roger, Friendship Seven, this is Zanzibar Cap Com. Proceed with 30-minute report. Over.
00 31 18	3.2	P	Roger, this is Friendship Seven. Standby one.
00 31 34	2.2	P	This is Friendship Seven, blood pressure.

00 32 04	11.7	S	Friendship Seven, Friendship Seven, this is Surgeon Zanzibar. You've got a good blood pressure trace. It shows the systolic and diastolic, if there's no necessity to repeat, you do not need to.
00 32 15	3.1	P	Roger, this is Friendship Seven, going through exercise.
00 32 48	3.2	P	This is Friendship Seven. Exercise completed, repeating blood pressure.
00 32 54	14.1	S	Friendship Seven, this Surgeon Zanzibar. Your blood pressure was 136 over 80 before exercise. We have good electrocardiographic trace during the time of exercise and you are now in a good level coming down on your blood pressure. Over.
00 33 08	1.9	P	Roger, this is Friendship Seven.
00 33 23	20.7	S	Friendship Seven, this is Zanzibar Surgeon. Blood pressure 136 systolic after exercise, recording well and coming down now to just under 90 for diastolic. Both traces are of excellent quality. Your electrocardiogram is excellent also. Everything on the dials indicates excellent aeromedical status. Over.
00 33 43	3.9	P	Roger, Friendship Seven. Running through 30-minute check,
00 33 52	23.3	P	This is Friendship Seven. The head movements caused no sensations, whatsoever. Feel fine. Reach test, I can hit directly to any spot that I want to hit. I have no problem reaching for knobs and have adjusted to zero-g very easily, much easier than I really thought I would. I have excellent vision of the charts, no astigmatism or any malfunctions at all.
00 34 15	0.9		. . .
00 34 20	1.9	P	Roger. You should get it now, okay.
00 34 23	0.9	CC	Now, thank you.
00 34 26	12.9	S	Friendship Seven, this is Zanzibar Surgeon, received your report indicating good reach accuracy. No disturbances on head motion, good visual acuity including astigmatism test and good response to exercise. Over. Could you ?
00 34 39	3.9	P	Roger, this is Friendship Seven. Are you ready to copy panel rundown? Over.
00 34 44	3.1	CC	Roger, Friendship Seven. This is Zanzibar CapCom. Proceed.
00 34 48	47.5	P	Ah, Roger, Zanzibar CapCom. Friendship Seven. All fuses remain same as previously reported; have not changed any of them. Squib is armed. Auto Retrojettison is off. ASCS is on normal, auto, gyro normal. All "T" handles are in. Retro Delay is normal. The sequence panel is normal. Landing bag is off. Fuel is 90-98 [percent]. The EPI is indicating just about right on schedule. My attitude is 5 [degrees] left, 3 [degrees] right, - 33 [degrees] on pitch. Retrograde time is still set for 04 plus 32 plus 28. Are you receiving? Over.
00 35 37	3.0	CC	Roger, Friendship Seven. Continue with the report.
00 35 40	46.9	P	Roger. This is Friendship Seven. The window, attitude indications, and periscope all check right together in good shape. I can see the dark side coming up in the periscope back behind me at present time. Cabin pressure is 5.5 and holding. Cabin temperature is 95 [degrees]. Relative humidity is 28 [percent]. I have turned the cabin — my suit temperature onto the increased water position for more cooling. Steam temperature is presently indicating 61 [degrees]. Oxygen is 75-100 [percent]. I didn't give suit temperature. Suit inlet temperature is 65 [degrees] and pressure is 5.8. Over.
00 36 28	2.7	CC	Roger, Friendship Seven. Continue with the report.
00 36 31	5.8	P	Roger. All other switches on right panel are normal except for Retrojettison and Retromanual fuse switches in the off position. Over.
00 36 42	3.4	CC	Roger, Friendship Seven. Could you give me suit exhaust temperature? Over.
00 36 45	6.5	P	Roger. Suit exhaust temperature, steam temperature is 61 [degrees]. I have just turned it down. Over.
00 36 54	4.3	CC	Roger, Friendship Seven. Continue with battery voltages. Over.
00 36 58	20.2	P	Roger. Battery voltages: Main is 24, Number One is 25, Two is 25, Three is 25, Standby One 25, Standby Two 25, Isolated 29. Ammeter is 22; ASCS, 112; fans, 113. Over.

00 37 20	7.5	CC	Roger, Friendship Seven. Are you ready to copy your Contingency Area I Delta retrosequence time? Over.
00 37 28	1.7	P	Roger. Say I Delta.
00 37 30	4.8	CC	I Delta is 00 50 24. Did you copy? Over.
00 37 35	4.0	P	Roger. 00 50 24 for I Delta.
00 37 40	9.4	CC	Roger, your inverter temperatures 150 volt inverter is 188°, 250 volt inverter, 158°.
00 37 50	4.0	P	Roger, Roger, I copy inverter temperatures okay.
00 37 59	12.3	CC	Friendship Seven, this is Zanzibar Cap Com. telemetry indicates all systems go. Medical status go.... reports same, we have a little bit of telemetry drop out .... Standby to pick up IOS in about 3 minutes. Zanzibar Cap Com out.
00 38 22	1.0	P	Roger, Zanzibar.

INDIAN OCEAN SHIP (FIRST ORBIT)

00 38 26	7.1	CT	Hello, Friendship Seven, Friendship Seven. This is Indian Com Tech ....
00 38 41	2.8	P	This is Friendship Seven, going on to eye patch.
00 39 04	4.3	CT	Hello, Friendship Seven. Friendship Seven, IOS Com Tech. Over.
00 39 09	2.9	P	Hello, IOS Com Tech, Friendship Seven. Go ahead.
00 39 28	5.0	CT	Friendship Seven, Friendship Seven. This is Indian Com Tech on HF UHF. How do you read? Over.
00 39 36	3.2	P	IOS, this is Friendship Seven. Do you read? Over.
00 39 49	5.2	CT	Friendship Seven, Friendship Seven. This is Indian Com Tech on HF UHF. How do you read? Over.
00 39 55	3.7	P	Indian Com Tech. Read you loud and clear on HF. Over.
00 40 08	5.5	CT	Hello Friendship Seven, Friendship Seven. This is Indian Com Tech on HF/UHF. How do you read? Over.
00 40 16	2.8	P	Indian Com Tech, Friendship Seven. Loud and clear, how me?
00 40 19	6.8	CT	Roger, Friendship Seven, reading you loud and clear
00 40 28	9.8	P	This is Friendship Seven. Had a beautiful sunset and can see the light way out almost up to the northern horizon.
00 40 38	1.6	CC	.....
00 40 57	5.0	CC	Friendship Seven, Friendship Seven, ...Over.
00 41 04	3.7	P	Indian Cap Com, I'm receiving you very garbled. Over.
00 41 09	2.7	CC	Roger, ... , now. Over.
00 41 44	3.6	CC	Friendship Seven, Friendship Seven, this is Indian Cap. Over.
00 41 48	3.0	P	Go ahead Indian Cap Com. I read you fairly good now. Over.
00 41 51	2.6	CC	Roger. Proceed with the rest of your status please. Over.
00 41 55	1.2	P	Say again, Indian. Over.
00 41 57	3.4	CC	Would you give me your status and consumable readings please? Over.
00 42 01	16.7	P	Roger. This is Friendship Seven. My status is very good. I feel fine. Fuel is 90-98 [percent], oxygen 75-100 [percent], amps, 21 present time, cabin pressure holding at 5.5. Over.
00 42 20	2.9	CC	Roger, understand, reading you loud and clear now. Over.
00 42 23	1.7	P	Roger, loud and clear.
00 42 31	30.1	P	This is Friendship Seven. At this, MARK, at this present time, I still have some clouds visible below me, the sunset was beautiful. It went down very rapidly. I still have a brilliant blue band clear across the horizon almost covering my whole window. The redness of the sunset I can still see through some of the clouds way over to the left of my course. Over.
00 43 03	1.0	CC	Roger, under ....
00 43 06	9.5	P	The sky above is absolutely black, completely black. I can see stars though up above. I do not have any of the constellations identified as yet. Over.
00 43 17	1.9	CC	Roger, understand Friendship Seven.
00 43 27	5.1	CC	Friendship Seven. Would you confirm you received retrosequence time I

			Delta from Zanzibar? Over.
00 43 34	5.0	P	This is Friendship Seven. Roger, I Delta is 00 50 24.
00 43 42	5.6	CC	Roger. I have areas I Echo, Foxtrot, and Hotel. Are you prepared to copy? Over.
00 43 47	1.1	P	Standby one.
00 43 49	1.1	CC	Roger. Let me know.
00 43 54	2.1	P	All right, go ahead with retrotimes. Over.
00 43 57	10.2	CC	Roger. Area I Echo is one hour, 15 minutes, 42 seconds. I say again, one hour, 15 minutes, 42 seconds. Over.
00 44 07	4.0	P	Roger. One Echo is 01 plus 15 plus 42.
00 44 33	9.3	CC	Roger. Area Foxtrot is one hour, 28 minutes, 50 seconds. I say again, one hour, 28 minutes, 50 seconds.
00 44 22	3.5	P	Roger. Foxtrot is 01 plus 28 plus 50.
00 44 26	9.4	CC	That is affirmative. Area Hotel is four hours, 32 minutes, 42 seconds. Four hours, 32 minutes, 42 seconds. Over.
00 44 36	3.2	P	Roger. 04 plus 32 plus 42. Over.
00 44 40	1.5	CC	Roger. That is affirmative. Over.
00 44 42	17.0	P	Roger. This is Friendship Seven. I am having no trouble at all seeing the night horizon. I think the moon is probably coming up behind me. Yes, I can see it in the scope back here and it's making a very white light on the clouds below. Over.
00 45 01	1.2	CC	Roger, understand.
00 45 08	3.9	CC	Friendship Seven, we have launched our flare. You understand it is … Over.
00 45 12	1.0	P	That is affirmative.
00 45 20	2.9	CC	We have been advised that it has been ignited. Do you see anything? Over.
00 45 25	5.8	P	This is Friendship Seven. Negative. I don't have anything in the scope or out the window.
00 45 35	0.9	CC	Roger, understand.
00 45 49	5.2	CC	Friendship Seven, this is Indian Cap Com. Do you have any feelings from weightlessness? Over.
00 45 54	3.8	P	This is Friendship Seven. Negative, I feel fine so far. Over.
00 45 59	0.8	CC	Roger, understand.
00 46 03	8.0	P	This is Friendship Seven. Turning suit water to increase position. I am still running steam temperature of about 60 [degrees]. Over.
00 46 13	0.8	CC	Roger, understand.
00 46 15	5.3	P	I am now on suit temperature setting of 1.7. Over.
00 46 21	3.1	CC	Roger, understand. What is your control mode? Over.
00 46 24	2.9	P	Control mode is ASCS automatic. Over.
00 46 28	0.7	CC	Roger, understand.
00 46 32	5.7	P	Roger. The night side is light enough; I can even see the horizon okay out through the periscope. Over.
00 46 40	1.1	CC	Roger. Over.
00 46 42	1.2	P	Friendship Seven, Roger.
00 46 59	3.2	CC	Friendship Seven, this is Indiana CapCom. We now have IOS. Over.
00 47 02	1.6	P	Roger, understand IOS.
00 47 08	2.5	CC	Your voice transmissions are starting to fade very badly. Over.
00 47 11	1.3	P	Roger. Friendship Seven.
00 47 15	2.1	CC	… clear, IOS Cap Com. Out.
00 47 18	0.5	P	Roger.
00 48 55	43.7	P	This is Friendship Seven, broadcasting in the blind. Wait a minute. Friendship Seven, broadcasting in the blind, making observations on night outside. There seems to be a high layer way up above the horizon; much higher than anything I saw on the daylight side. The stars seem to go through it and then go on down toward the real horizon. It would appear to be possibly some 7

or 8 degrees wide. I can see the clouds down below it; then a dark band, then a lighter band that the stars shine right through as they come down toward the horizon. I can identify Aries and Triangulum.

MUCHEA (FIRST ORBIT)

00 49 49	3.7	CT	Friendship Seven, Muchea Com Tech. We read you. Would you.
00 49 55	4.8	P	Hello, Muchea, Com Tech. This is Friendship Seven, reading you loud and clear. How me?
00 50 01	3.1	CC	Roger, Friendship Seven. Muchea Cap Com. How me? Over.
00 50 05	5.1	P	Roger. How are you doing Gordo? We're doing real fine up here. Everything is going very well. Over.
00 50 10	1.2	CC	John, you sound good.
00 50 12	12.4	P	Roger. Control fuel is 90-100 [percent], oxygen is 75-100 [percent], amps are 22, all systems are still go. Having no problems at all. Control system operating fine. Over.
00 50 25	3.6	CC	Roger. Do you have any star or weather or landmark observations as yet? Over.
00 50 29	27.4	P	Roger. I was just making some to the recorder. The only unusual thing I have noticed is a rather high, what would appear to be a haze layer up some 7 or 8 degrees above the horizon on the night side. The stars I can see through it as they go down toward the real horizon, but it is a very visible single band or layer pretty well up above the normal horizon. Over.
00 50 58	1.2	CC	Roger, very interesting.
00 51 00	15.5	P	This is Friendship Seven. I had a lot of cloud cover coming off of Africa. It has thinned out considerably now and although I can't definitely see the ocean, there is a lot of moonlight here that does reflect off what clouds there are, Over.
00 51 16	12.4	CC	Roger. You had an excellent cutoff, John. Your velocity was 8 feet per second low. V over Vr, 1.0002. Can you confirm your retrosequence I Easy, Foxtrot, and Hotel from Indian Ocean Ship? Over.
00 51 29	1.4	P	That is affirmative. I did.
00 51 32	6.1	CC	Roger. Your yaw check over ATS was good. Your yaw when you were on fly by-wire over IOS was excellent.
00 51 38	0.5	P	Roger.
00 51 40	2.5	CC	Are we clear to send you a Z and R Cal? Over.
00 51 44	1.0	P	Roger.
00 51 46	1.3	CC	Z Cal coming through now.
00 51 51	5.0	CC	Roger. Shortly you may observe some lights down there. You want to take a check on that to your right? Over.
00 51 56	3.5	P	Roger. I'm all set to see if I can't get them in sight.
00 52 00	3.4	CC	Roger. You do have your visor closed at this time. Over.
00 52 03	5.6	P	This is affirmative. I had it open for a little while; it's closed now. Cabin pressure is holding in good shape. Over.
00 52 10	2.7	CC	Roger, Z Cal is off. R Cal is coming through now.
00 52 13	0.5	P	Roger.
00 52 16	2.9	CC	Any symptoms of vertigo or nausea at all? Over.
00 52 19	3.0	P	Negative, no symptoms whatsoever. I feel fine. Over.
00 52 23	0.4	CC	Very well.
00 52 31	5.2	CC	Roger. Your 150 VA inverter is 180°. Looks like it's doing pretty well.
00 52 36	2.2	P	Roger. Looks like it is holding up fine.
00 52 46	0.8	CC	R Cal is off.
00 52 48	0.4	P	Roger.
00 53 01	1.3	P	That was sure a short day.
00 53 04	1.2	CC	Say again, Friendship Seven.

00 53 05	2.6	P	That was about the shortest day I've ever run into.
00 53 08	1.6	CC	Kinda passes rapidly, huh.
00 53 10	0.4	P	Yes Sir.
00 53 12	0.3	CC	Fine.
00 53 15	5.1	CC	Okay. Do you have any landmark, any other landmark observations to make? Over.
00 53 20	10.6	P	This is Friendship Seven. I have the Pleiades in sight out here, very clear. Picking up some of these star patterns now. Little better than I was just off of Africa.
00 53 33	3.5	CC	Roger, understand you have Pleiades in sight. Have you sighted Orion yet? Over.
00 53 38	2.4	P	Negative. Do not have Orion in sight yet.
00 53 41	7.6	CC	Within a few seconds, you should have Orion and Canopus and Sirius probably in sight very shortly thereafter. The moon will be off....
00 53 50	0.4	P	Roger.
00 53 56	2.4	CC	Do you have time to send us a blood pressure reading? Over.
00 53 59	1.5	P	Roger. Standby.
00 54 03	0.3	CC	Roger.
00 54 20	3.2	CC	Roger. The surgeons are standing by for your blood pressure whenever you're ready.
00 54 25	2.5	P	Roger. I'm already sending it. Did they pick it up? Over.
00 54 28	1.5	CC	Roger. They have it in good shape.
00 54 31	4.8	P	Roger. I do have the lights in sight on the ground. Over.
00 54 36	2.5	CC	Roger. Is it just off to your right there?
00 54 39	12.3	P	That's affirmative. Just to my right I can see a big pattern of lights apparently right on the coast. I can see the outline of a town and a very bright light just to the south of it. On down ....
00 54 52	2.1	CC	Perth and Rockingham, you're seeing there.
00 54 54	4.9	P	Roger. The lights show up very well and thank everybody for turning them on, will you?
00 55 00	0.9	CC	We sure will, John.
00 55 02	17.2	P	Very fine. On down farther to the south and inland, I can see more lights. There are two, actually four patterns in that area. And also, coming into sight in the window now is another one almost down under me. The lights are very clear from up here.
00 55 19	1.3	CC	Roger. Sounds good.

WOOMERA (FIRST ORBIT)

00 55 26	6.8	CC	Friendship Seven, Friendship Seven. This is Woomera Cap Com, reading you loud and clear. We have TM solid. Woomera standing by.
00 55 33	1.3	P	Roger, Woomera.
00 55 40	8.2	CC	Friendship Seven. We have your blood pressure readout, reading 126 over 90. What kind of results from your physiological tests? Over.
00 55 49	20.7	P	This is Friendship Seven. I have had no ill effects at all as yet from any zero-g. It's very pleasant, in fact. Visual acuity is still excellent. No astigmatic effects. Head movements caused no nausea or discomfort whatsoever. Over.
00 56 11	6.1	CC	Roger, Friendship Seven. Let's go ahead with this 30-minute report please, starting with the fuse panel positions. Over.
00 56 17	46.1	P	Roger. This is 30-minute report coming up. Have gone through head movements and I get no effect from that. Have gone through the reach test and have no problem with that at all. My orientation is good, vision is clear. My moving target, looking at a light spot back and forth causes no ill effects whatsoever. Running the light test at present time and all lights do check okay, in the capsule. I am over a large cloud bank at present time, almost

			extending to the left of my course which would be to the south. Over.
00 57 04	3.2	CC	Roger, Friendship Seven. Let's continue on with this 30-minute report.
00 57 07	1:13.9	P	Roger. Panel rundown follows. The fuse switches on the left panel are all as previously reported, without repeating them again. Squib is armed. Auto Retro jettison is off. ASCS Mode Select is normal, auto, gyro normal. All "T" Handles are in. The Retro, Delay is normal, Cabin lights are on both. I have the red filters on, of course. The Telem low frequence is on. Rescue Aids is auto. All sequence panel items are normal except the Landing Bag to the center off position. Fuel is 90-100 [percent]. Attitude indications are 5 [degrees] right, 3 [degrees] left, 36 [degrees] down. EPI is indicating fairly well, indicating almost over Woomera at present time. Time from launch on my mark will be 58 plus 15. Standby-MARK. Retrograde time is still set at 04 plus 32 plus 28.
00 58 23	34.6	P	Cabin pressure is holding steady at 5.5. Temperature in the cabin is 100 [degrees]. Cabin humidity is 25 percent, Coolant quantity is 67 percent. Suit environment is 62 [degrees]. Inlet pressure. Inlet temperature. . . . pressure is 5.8 on the suit. Steam temperature is 57 [degrees] and dropping since I turned it down. Oxygen is 74-101 [percent]. Amps 23. Are you still copying? Over.
00 58 59	1.7	CC	Affirmative, Friendship Seven. Go ahead.
00 59 07	27.3	P	Roger. Main Bus is 24, One is 25, 25, 25. Standby One is 25, 25. Isolated 29, and Main 24 again. Fans 113, ASCS 113. All switch fuses on the right are on except Retrojettison and Retromanual in the center-off position. All other switches normal. Over.
00 59 32	11.7	CC	Hello. Roger, Friendship Seven. Read you loud and clear. All systems appear go at this time. Your clocks are in sync. We have the Woomera Airport lights on. Do you, do you see? Over.
00 59 44	12.9	P	This is Friendship Seven. Negative, I do not; there's too much cloud cover in this area. I had the lights at Perth in good shape; they were very clear, but do not have the lights at Woomera; sorry. Over.
00 59 58	1.3	CC	Roger, Friendship Seven.
01 00 01	2.3	CC	Friendship Seven, let's have another blood pressure check please. Over.
01 00 05	1.4	P	Friendship Seven, Roger.
01 00 24	4.5	CC	Hello, Friendship Seven, Woomera Cap Com. 150 volt inverter temperature reading 195 [degrees]; 250 inverter temperature reading 172 degrees.
01 00 34	0.4	P	Roger.
01 00 45	3.1	CC	Friendship Seven, this is Woomera Cap Com. What's your cabin temperature reading? Over.
01 00 50	2.1	P	Woomera, standby just a moment, please.
01 00 58	1.9	P	Say again. You wanted cabin temperature?
01 01 01	0.9	CC	Roger, Friendship Seven.
01 01 02	1.8	P	Cabin temperature 100 [degrees]. Over.
01 01 06	6.1	CC	Roger, Friendship Seven. Your Roll Auto Line temperature is reading 110 [degrees].
01 01 12	1.5	P	Roger.
01 01 14	1.7	CC	Temperature is reading 95 [degrees]. Over.
01 01 16	1.5	P	Roger, Friendship Seven.
01 01 19	4.5	P	This is Friendship Seven. Getting some pictures of star. Over.
01 01 26	1.1	CC	Roger, Friendship Seven.
01 02 00	2.6	CC	Hello Friendship Seven, Woomera Cap Com. Do you read me? Over.
01 02 04	3.2	P	Roger, Woomera Cap Com. Still read you loud and clear. How me?
01 02 08	14.3	CC	Weak but readable, Friendship Seven. All systems appear go at this time. Blood pressure reading 125 over 90. You have an HF check coming up in about 2 minutes and 30 seconds, and Canton acquisition approximately 7 minutes. Over.

01 02 22	8.0	P	Roger, Friendship Seven. Thank you very much. All systems are still go at this time. Switching to HF. Over.
01 02 31	2.2	CC	Roger, Friendship Seven. This is Woomera Cap Com. Out.
01 03 17	3.1	P	Hello Woomera, Friendship Seven on HF. Do you receive me? Over.
01 04 00	13.5	P	This is Friendship Seven. This is Friendship Seven, broadcasting in the blind to the Mercury network. 1, 2, 3, 4, 5. This is Mercury Friendship Seven. Out.
01 04 16	11.9	CC	This is Canaveral Cap Com, testing the HF. G.m.t. 15 (Cape) 52 05, MARK. I did not read the capsule. Cape, out.
01 05 17	10.0	CC	Friendship Seven, this Kano. G.m.t. 15 53 02. I did (Kano) not read your transmission. Kano, out.
01 05 32	7.8	CC	Friendship Seven, this is Zanzibar Cap Com. . . . This is Zanzibar Cap Com, out.

CANTON (FIRST ORBIT)

01 09 44	2.7	CT	Friendship Seven. This is Canton Com Tech. Do you read? Over.
01 09 53	4.9	P	Hello, Canton. Hello, Canton. This is Friendship Seven. Read you loud and clear, how me? Over.
01 10 00	3.9	P	Hello Canton. Hello Canton, Friendship Seven. Read you loud and clear. How me? Over.
01 10 08	5.1	CT	Friendship Seven, Canton Com Tech. Read you loud and clear, also. Stand by for Cap Com.
01 10 14	13.8	P	Roger. This is Friendship Seven. Com Tech Canton. Fuel 90-100 [percent], oxygen 73-101 [percent].
01 10 28	2.4	CC	Seven, this is Canton Cap Com. Go ahead. Over.
01 10 32	15. 1	P	Roger, Canton Cap Com. My report follows: Fuel 90-100 [percent], oxygen 73-100 [percent], amps 22, cabin pressure holding steady at 5.5 Over.
01 10 50	4.0	CC	Friendship Seven, this is Canton Cap Com. Would you repeat that please?
01 10 55	5.6	P	Roger, Canton. Friendship Seven repeating. First, do you have TM solid? Over.
01 11 02	4.1	CC	Friendship Seven, this is Canton. Affirmative, we do have TM solid.
01 31 07	12.7	P	Roger, message follows: Control fuel 90-100 [percent], oxygen 73-100 [percent], amps 22, cabin pressure 5.5 and holding steady. Over.
01 11 21	4.2	CC	All, Roger, Friendship Seven. What control mode are you in the present time? Over.
01 11 26	4. 2	P	This is Friendship Seven. Still in automatic, automatic. Over.
01 11 32	2.8	CC	Roger, Friendship Seven. Canton standing by.
01 11 35	1.4	P	Roger, Friendship Seven.
01 12 06	17.0	CC	Friendship Seven. This is Canton Cap Com. All systems on the ground appear go. Your 150 volt-amp inverter temperature is presently 188°, repeat 188°. Your clocks are in sync. Over.
01 12 23	1.8	P	Roger, Friendship Seven. Thank you.
01 12 29	2.7	P	This is Friendship Seven, preparing to eat. Over.
01 12 35	1.4	CC	Roger, Friendship Seven.
01 12 42	9.6	P	This is Friendship Seven. I had a big storm in sight off to the south, of course, for a little while and had lightning flashes going around on top of the clouds. I could see it very clearly. Over.
01 12 53	0.3	CC	Roger.
01 13 09	3.2	P	This is Friendship Seven. Opening visor, going to eat. Over.
01 13 13	0.9	CC	Roger, Friendship Seven.
01 14 60	13.5	P	This is Friendship Seven. Having no trouble at all eating, very good. In the periscope, I can see the brilliant blue horizon coming up behind me; approaching sunrise. Over.
01 14 15	2.3	CC	Roger, Friendship Seven. You are very lucky.
01 14 19	1.6	P	You're right. Man, this is beautiful.

01 14 30	16.0	P	This is Friendship Seven. Have eaten one tube of food, shutting the visor. I've had no problem at all eating. Oh, the sun is coming up behind me in the periscope, a brilliant, brilliant red. Over.
01 14 48	0.4	CC	Roger.
01 14 53	12.2	P	This is Friendship Seven. It's blinding through the scope on clear. It's started up just as I gave you that mark; I'm going to the dark filter to watch it come on up.
01 15 06	0.4	CC	Roger.
01 15 24	21.3	P	This is Friendship Seven. I'll try to describe what I'm in here. I am in a big mass of some very small particles, that are brilliantly lit up like they're luminescent. I never saw anything like it. They round a little; they're coming by the capsule, and they look like little stars. A whole shower of them coming by.
01 15 57	12.6	P	They swirl around the capsule and go in front of the window and they're all brilliantly lighted. They probably average maybe 7 or 8 feet apart, but I can see them all down below me, also.
01 16 06	3.1	CC	Roger, Friendship Seven. Can you hear any impact with the capsule? Over.
01 16 10	16.8	P	Negative, negative. They're very slow; they're not going away from me more than maybe 3 or 4 miles per hour. They're going at the same speed I am approximately. They're only very slightly under my speed. Over.
01 16 33	10.1	P	They do, they do have a different motion, though, from me because they swirl around the capsule and then depart back the way I am looking.
01 16 46	1.3	P	Are you receiving? Over.
01 16 55	2.5	P	There are literally thousands of them.
01 17 16	3.7	P	This is Friendship Seven. Am I in contact with anyone? Over.
01 17 30	11.7	P	This has been going on since about I plus 15. Over. Just after I remarked about the sunset. I looked back up and looked out the window, and all the little swirl of particles was going by. Over.
01 19 24	6.6	P	This is Friendship Seven. This is Friendship Seven, broadcasting in the blind again on HF.
01 19 38	40.9	P	This is Friendship Seven, broadcasting in the blind. Sunrise has come up behind in the periscope. It was brilliant in the scope, a brilliant red as it approached the horizon and came up; and just as the — as I looked back up out the window, I had literally thousands of small, luminous particles swirling around the capsule and going away from me at maybe 3 to 5 miles per hour. Now that I am out in the bright sun, they seem to have disappeared. It was just as the sun was coming up. I can still see just a few of them now, even though the sun is up some 20° above the horizon.

GUAYMAS (FIRST ORBIT)

01 20 44	7. 1	CT	Friendship Seven, Friendship Seven. This is Guaymas Com Tech, Guaymas Com Tech, transmitting on HF-UHF. Do you read? Over.
01 20 51	7.8	P	Hello, Guaymas Com Tech, Guaymas Core Tech. This is Friendship Seven. Receive your HF loud and clear. How me? Over.
01 21 00	9.0	CC	Friendship Seven, Friendship Seven. This is Guaymas Cap Com. You're little garbled right now, but understandable. What is, what is your status? Over.
01 21 08	15.6	P	This is Friendship Seven on HF. My status is excellent. Everything is going according to plan. Control fuel is 90-100 [percent], oxygen is 72-101 [percent], amps 22. Over.
01 21 29	4.3	CC	Roger, Friendship Seven. Would you repeat your oxygen please? Over.
01 21 34	19	P	This is Friendship Seven. Oxygen is 72-101 [percent]. Over.
01 21 42	0.6	CC	Roger.
01 21 50	4.9	CC	Friendship Seven, Guaymas Cap Com. Everything looks fine from here.
01 21 57	3.9	P	Roger, This is Friendship Seven. Understand everything looks okay.

01 22 01	6.3	CC	. . . 110. Your main inverter temperatures are rising slightly, but everything is okay now. Over.
01 22 09	1.7	P	Roger. Friendship Seven.
01 22 32	6.4	P	This is Friendship Seven. I still have some of these very small particles coming around the capsule. Over.
01 22 42	3.2	CC	Friendship Seven, Guaymas Cap Com. Say again, please. Over.
01 22 46	17.3	P	This is Friendship Seven. Just as the sun came up, there were some brilliantly lighted particles that looked luminous, that were swirling around the capsule. I don't have any in sight right now; I did have a couple just a moment ago, when I made the transmission to you. Over.
01 23 10	0.9	CC	Roger, Friendship Seven.
01 23 21	9.5	P	This is Friendship Seven. For the record, number, the number two film that I'm putting in the camera is the number four roll.
01 23 52	19.1	CC	Friendship Seven. Guaymas Cap Com. We have about 3 minutes yet before we acquire telemetry. I'll give you, I'll give you your retrosequence times for Areas 2 Alpha, Golf, Hotel. Area 2 Alpha is 01 36 38.
01 24 11	6.2	P	This is Friendship Seven. Standby. I'm not ready to copy yet; I'm changing film in the camera. Over.
01 24 23	1.4	CC	Roger, we're standing by.
01 24 52	6.1	P	This is Friendship Seven, going back and using the number 2 roll if I can get hold of it.
01 25 07	3.1	P	Negative, number 2 is missing; I'm using number 3.
01 25 17	2.6	CC	Friendship Seven, Guaymas Cap Com. Say again, please.
01 25 21	7.1	P	This is Friendship Seven, I was making a transmission for the record here on what film I was loading. Over.
01 25 31	0.4	CC	Roger.
01 26 05	4.1	CC	Friendship Seven, Guaymas Cap Com. Could you go to UHF now?
01 26 10		P	Roger. This is Friendship Seven, going to UHF.
01 26 40	3.2	P	Hello, Guaymas, Friendship Seven on UHF. Do you receive? Over.
01 26 44	1.8	CC	Roger, Friendship Seven, loud and clear. Over.
01 26 47	5.3	P	Hello, Roger, this is Friendship Seven. Loud and clear with you also. Are we still go from Control Center? Over.
01 26 53	1.0	CC	Roger, we're still go.
01 26 55	12.3	P	Roger. Friendship Seven. All systems are go in the capsule. I still have some of these little particles coming around the capsule occasionally here. I can see them against the dark sky even on the day side. Over.
01 27 10	1.3	CC	Roger, understand.
01 27 19	5.9	CC	Friendship Seven, Guaymas. Do you want your contingency times now? Over.
01 27 27	2.3	P	This is Friendship Seven. Ready to copy.
01 27 32	S. 5	CC	Area 2 Alpha 01 36 38. Area Golf 03 00 41.
01 27 42	2.1	P	Say again Area Golf again, please.
01 27 45	4.3	CC	Area Golf 03 00 41.
01 27 50	4.6	P	Roger, Area Golf is 03 plus 00 plus 41. Is that affirm?
01 27 56	5.3	CC	That's affirmative. Area Hotel is 04 32 40.
01 28 02	4.5	P	Roger, 04 plus 32 plus 40 for Hotel.
01 28 13	5.5	CC	Friendship Seven, Guaymas. If you have a chance, could you give us a blood pressure check? Over.
01 28 20	8.0	P	Roger. Friendship Seven. Will give blood pressure check. I still have some of these particles that I cannot identify coming around the capsule occasionally. Over.
01 28 30	1.7	CC	Roger. How big are these particles?
01 28 32	16.7	P	Very small, I would indicate they are on the order of 16th of an inch or smaller. They drift by the window and I can see them against the dark sky. Just at sunrise there were literally thousands of them. It looked just, like a myriad of stars. Over.

01 28 50	3.8	CC	Roger. Are they moving by you or floating with you? Over.
01 28 54	9.2	P	Some of them almost float with me. Most of them appear to be moving at about 3 to 5 miles an hour away from me. I'm going just a little faster than they are. Over.
01 29 05	0.5	CC	Roger.
01 29 44	6.4	CC	Friendship Seven, Guaymas. Cap Com. We have a scope retract indication on ground. Believe it's ground failure. Over.
01 29 52	3.8	P	That is affirmative. The scope is extended. The scope is extended. Over.
01 29 57	0.4	CC	Roger.
01 30 03	3.7	P	This is Friendship Seven. A lot of cloud cover. I cannot see land yet. Over.
01 30 11	11.5	P	This is Friendship Seven. Yaw is drifting out of orbit attitude and will bring it back in. Over.
01 30 16	0.4	CC	Roger.
01 30 18	7.0	CC	Friendship Seven, Guaymas. I said the, I have an indication of scope retract. You say your scope is out, though? Over.
01 30 26	0.8	P	That is affirmative.
01 30 28	0.4	CC	Roger.

CALIFORNIA (FIRST ORBIT)

01 30 30	3.7	CC	Friendship Seven, California Cap Com. We have your scope as extended.
01 30 34	2.7	P	That is affirmative, Cal, scope is extended.
01 30 39	1.3	CC	Roger, we concur here.
01 30 41	0.4	P	Roger.
01 30 44	9.1	P	This is Friendship Seven. Yaw drifted out of limits about 20 degrees to the right. I'm bringing it back in manually at present time. Over.
01 30 55	0.5	CC	Roger.
01 31 03	2.6	p	This is Friendship Seven. Back on ASCS.
01 31 15	3.9	CC	Friendship Seven, Guaymas. You were on ASCS when it started drifting out. Over.
01 31 19	1.2	P	That is affirmative.
01 31 25	4.8	P	This is Friendship Seven. Now on fly-by-wire to hold orbit attitude.
01 31 33	6.5	p	This is Friendship Seven. I have the land in sight out the window. Controlling manually, on fly-by-wire.
01 31 41	3.0	CC	Friendship Seven. Is the ASCS OK now? Over.
01 31 45	2.5	P	I'll switch back on to it now and see.
01 32 03	3.8	P	This is Friendship Seven. Negative. Seems to, well, standby one.
01 32 15	5.1	P	This is Friendship Seven. Affirmative. Seems to be controlling now. There may be some drift to the right, however. Over.
01 32 22	0.4	CC	Roger.
01 32 29	17.3	p	This is Friendship Seven. It drifts about out 20° to the right and then gets a large pulse to kick it back over to the left. It goes in about, the rate runs over to about 3 degrees per second, to the left.
01 32 45	4.3	p	It's drifting again in yaw. Over. And once again pulls through.

CANAVERAL (SECOND ORBIT)

01 32 50	1.5	CT	Cape Com Tech. Do you read? Over.
01 32 52	2.3	P	Roger, Cape Com Tech. Friendship Seven. Over.
01 33 03	2.3	P	Hello, Cape Com Tech. Friendship Seven. Over.
01 33 14	5.1	CT	Friendship Seven, this is Cape Com Tech, Cape Com Tech. Do you read? Over.
01 33 19	2.2	P	Roger, Cape Com Tech. Friendship Seven.
01 33 28	2.0	P	Hello, Cape Com Tech. Friendship Seven. Over.
01 33 27	3.8	CT	Friendship Seven, this is Canaveral Com Tech. How do you copy? Over.

01 33 31	4.9	P	Friendship Seven to Canaveral. Read you loud and clear. How me? Over.
01 33 37	4.0	CT	Friendship Seven, Canaveral Com Tech. Read you loud and clear. Standby for Cap Com. Over.
01 33 42	0.4	P	Roger.
01 33 44	7.4	CC	Friendship Seven, Cap Com. Would you give us the difficulty you've been having in yaw in ASCS? Over.
01 33 52	39.0	P	Roger, This is Friendship Seven. I'm going on fly-by-wire so I can control more accurately. It just started as I got to Guaymas, and appears to be it drifts off in yaw, to the right at about 1° per second. It will go over to an attitude of about 20°, and hold at that and when it hits about a 20° point it then goes into orientation mode and comes back to zero, and it was cycling back and forth in that mode. I am on fly-by-wire now and controlling manually. Over.
01 34 32	5.4	CC	Roger. Understand. Do you have a retrofire time for 2 Bravo and 2 Charlie?
01 34 39	1.9	P	This is Friendship Seven. Negative.
01 34 41	12.2	CC	OK, 2 Bravo, 01 50 00; 2 Charlie, 02 05 59. Over.
01 34 57	20.6	P	Ah, this is Friendship Seven. Understand 1 Bravo is 01 plus 50 plus 00; 2 Charlie, correction 2 Bravo is 01 50 00, 2 Charlie is 02 plus 05 plus 59. Is that affirm?
01 35 09	2.1	CC	That is affirmative. Standby one, please.
01 35 27	13.2	P	This is Friendship Seven. What appears to have happened is, I believe, I have no one pound thrust in left yaw. So it drifts over out of limits and then hits it with the high thrust. Over.
01 35 41	3.8	CC	Roger, Seven, we concur here. Recommending you remain fly-by-wire.
01 35 45	1.8	P	Roger, remaining fly-by-wire.
01 35 56	7.0	CC	Seven, this is Cape. The President will be talking to you and while he is talking I'll be sending Z and R cal.
01 36 05	1.2	P	Ah-President.
01 36 07	0.9	CC	Go ahead, Mr. President.
01 36 47	1.7	P	This is Friendship Seven, standing by.
01 36 50	3.6	CC	Roger, Seven. Having a little difficulty. Start off with your 30 minute report.
01 36 54	39.4	P	Roger. This is Friendship Seven, controlling manually on fly-by-wire, having no trouble controlling. Very smooth and easy; controls very nicely. Fuses are still the same on the left panel. Squib is, Squib is armed. Auto Retrojettison is off. ASCS is fly-by-wire, auto, gyro normal. All fuel, all "T" handles are in, Retro Delay is normal. I have beautiful view out the window of the coast at present time. Just departing. Can see the, way down across Florida. Cannot quite see the Cape yet.
01 37 35	0.3	CC	Roger.
	22.4	P	Continuing with the report. No sequence panel lights showing. Only abnormal position. Landing bag is off. The EPI is indicating okay. Control fuel is 80 [percent] auto, 100 [percent] manual. Retrograde time set in is 04 plus 32 plus 28. Over.
01 38 00	1.3	CC	Roger, still reading you.
01 38 02	2.4	P	Roger. Are we in communication yet? Over.
01 38 06	1.0	CC	Say again, Seven.
01 38 08	5.1	P	Roger. I'll be out of communication fairly soon. I thought if the other call was in, I would stop the check. Over.
01 38 14	1.9	CC	Not as yet, we'll get you next time.
01 38 16	40.7	P	Roger. Continuing report. Cabin pressure 5.5 and holding nicely. Cabin air is 95 [degrees]; relative humidity, 20 [percent]; coolant quantity is 67 [percent]; temperature is 67 [degrees] on the suit; 5.8 on the pressure; steam temperature is 50 [degrees] and coming down slowly. Oxygen is 70-100 [percent]; amps, 22. Only really unusual thing so far beside ASCS trouble were the little particles, luminous particles around the capsule, just thousands

of them right at sunrise over the Pacific. Over.

01 38 59	5.2	CC	Roger, Seven, we have all that. Looks like you're in good shape. Remain on fly-by-wire for the moment.
01 39 04	1.3	P	Roger. Friendship Seven.

BERMUDA (SECOND ORBIT)

01 39 09	1.7	CC	Friendship Seven, this is Bermuda Cap Com.
01 39 11	0.7	P	Go ahead, Bermuda.
01 39 13	1.8	CC	Roger, we read you 5 by 5.
01 39 15	2.0	CC	Seven, this is Cape. Go to Bermuda now. (CNV)
01 39 18	2.6	P	Roger. This is Friendship Seven. Go ahead, Bermuda.
01 39 24	4.5	CC	Roger, Friendship Seven. Your oculogyric test is due in a minute and 30 seconds.
01 39 29	7.8	P	Roger. This is Friendship Seven. I am controlling fly-by-wire, present time. I have no left yaw low thrust. Over.
01 39 38	1.9	CC	Roger, we understand all of your reports.
01 39 40	0.4	P	Roger.
01 40 19	0.9	CC	Friendship Seven. Bermuda.
01 40 20	1.4	P	Go ahead, Bermuda. Friendship Seven.
01 40 23	2.1	CC	Have you started your oculogyric test yet?
01 40 25	2.2	P	Negative, not yet. Just getting set at the present time.
01 41 00	1.7	CC	Friendship Seven, have you started your test yet?
01 41 02	2.4	P	This Friendship Seven. That's affirmative, starting now.
01 41 05	0.5	CC	Roger.
01 41 50	4.0	P	This is Friendship Seven. I can get no, no results from that at all.
01 42 05	2.0	P	Hello, Bermuda Cap Com. Friendship Seven. Over.
01 42 09	0.4	CC	Friendship Seven.
01 42 09	4.4	P	This is Friendship Seven. Oculo check being completed. Got no effects.
01 42 13	0.8	CC	. . . Com. Go ahead.
01 42 14	5.2	P	Bermuda Cap Com, Friendship Seven. I get no results from oculo check.
01 42 20	1.3	CC	Cap Com. If you read, go ahead.
01 42 22	3.4	P	Bermuda, Friendship Seven. I get no results from oculo check. Over.
01 42 41	1.6	CC	Friendship Seven, Bermuda Cap Com on HF.
01 42 44	2.6	P	Bermuda, this is Friendship Seven. Do you receive me now? Over.
01 42 55	1.8	CC	Friendship Seven, Bermuda Cap Com. How do you read?
01 42 59	3.6	P	Hello, Bermuda Cap Com. Friendship Seven, HF. Do you receive me now? Over.
01 43 03	3.9	CC	On HF. What are the results of the oculogyric test, please?
01 43 07	3.7	P	Results negative. Get no effect at all, no effect at all. Over.
01 43 11	2.6	CC	Roger. Understand no effects from oculogyric. Thank you.
01 43 14	0.7	P	That's affirmative.
01 43 38	2.2	CC	Friendship Seven, Bermuda Cap Com. Do you read?
01 43 41	2.1	P	Roger, Bermuda. Loud and clear; how me?
01 43 44	1.4	CC	Roger. We still are reading you loud and clear.
01 43 50	3.4	P	Roger. Friendship Seven getting ready for I plus 44 check.
01 44 01	14.4	P	This is Friendship Seven broadcasting blind to Mercury Network on from just off the east coast over middle Atlantic, check completed at 15.
01 44 17	7.4 (CNV)	CC	. . . testing on HF G.m.t. 16 31 59 MARK. I read the capsule weak, but readable. Cape, out.
01 44 31	6.7	CC	Friendship Seven, this is Bermuda Cap Com on HF. We did not read the capsule. G.m.t. 16 32 15.

CANARY (SECOND ORBIT)

01 44 41	11.5	CC	Friendship Seven, Friendship Seven, this is CYI. The time now 16 32 26. We

			are reading you loud and clear; we are reading you loud and clear. CYI.
01 44 57	10.7	P	This is Friendship Seven on UHF. As I went over recovery area that time, I could see a wake, what appeared to be a long wake in the water. I imagine that's the ships in our recovery area.
01 45 09	5.8	CC	Friendship Seven, . . . We do not read you, do not read you. Over.
01 45 17	9.6 (KNO)	CC	Friendship Seven, this is Kano. At G.m.t. 16 33 00. We do not. . . . This is Kano. Out.
01 46 38	4.0	CT	Friendship Seven, Friendship Seven, this is CYI Com Tech. Over.
01 46 46	6.7	P	Hello, Canary. Friendship Seven. Receive you loud and a little garbled. Do you receive me? Over.
01 46 55	4.3	CT	Friendship Seven, Friendship Seven, this is CYI Com Tech. Over.
01 47 01	4.0	P	Hello, Canary, Friendship Seven. I read you loud and clear. How me? Over.
01 47 25	3.7	CT	Friendship Seven, Friendship Seven, this is CYI Com Tech. Over.
01 47 30	5.4	P	Hello, CYI Com Tech. Friendship Seven. How do you read me? Over.
01 47 48	4.2	CT	Friendship Seven, Friendship Seven, this is CYI, CYI Com Tech. Do you read? Over.
01 47 54	4.0	P	Roger. This is Friendship Seven, CYI. I read you loud and clear. Over.
01 48 15	4.9	CT	Friendship Seven, Friendship Seven, this is CYI Com Tech, CYI Com Tech. Do you read? Over.
01 48 20	2.9	P	Hello, CYI Com, Tech. Roger, read you loud and clear.
01 48 26	5.3	CT	Friendship Seven, this is CYI Com Tech. Read you loud and clear also, on UHF, on UHF. Standby.
01 48 32	1.3	P	Roger. Friendship Seven.
01 48 51	5.6	CC	Friendship Seven, Friendship Seven, Friendship Seven, this is Canary Cap Com. How do you read? Over.
01 48 58	3.3	P	Hello, Canary Cap Com. Friendship Seven. I read you loud and clear. How me?
01 49 02	11.9	CC	I read you loud and clear. I am instructed to ask you to correlate the actions of the particles surrounding your spacecraft with the actions of your control jets. Do you read? Over.
01 49 15	5.1	P	This is Friendship Seven. I did not read you clear. I read you loud but very garbled. Over.
01 49 22	15.5	CC	Roger. Cap asks you to correlate the actions of the particles surrounding the vehicle with the reaction of one of your control jets. Do you understand? Over.
01 49 39	4.1	P	This is Friendship Seven. I do not think they were from my control jets, negative. Over.
01 49 52	2.2	CC	Roger, I understand.
01 49 58	5.9	CC	Friendship Seven, Friendship Seven, this is Canary Cap Com. Please complete your status report.
01 50 06	8.7	P	This is Friendship Seven. My status is excellent. I have control of capsule on fly-by-wire at present time. Control fuel is 80-100 [percent]; oxygen, 69-100 [percent]; amps, 22. Over.
01 50 25	8.0	CC	Understand your auto fuel is 80 [percent], your manual fuel is 100 [percent], Main O_2 is 69 [percent], secondary O_2 is 100 [percent]. Over.
01 50 35	1.7	P	Roger. This is Friendship Seven.
01 50 40	2.2	CC	Friendship Seven, would you take a deep breath?
01 50 43	1.6	P	Roger. Friendship Seven. Deep breath.
01 50 50	0.5	CC	That's good.
01 50 52	6.3	P	This is Friendship Seven. Have Cape Verde Islands in sight to my left. Over.
01 50 59	1.0	CC	Roger, I understand.
01 51 04	2.0	CC	Friendship Seven, your medical status is excellent.
01 51 08	1.1	P	Roger. Friendship Seven.
01 51 15	6.4	P	This is Friendship Seven. The sun coming through the window is very warm where it hits the suit. I get quite a bit of heat from it.

01 51 23	0.5	CC	Roger, I understand.
01 51 28	3.8	CC	Friendship Seven, are you going through your day horizon check?
01 51 33	1.6	P	This is Friendship Seven. Say again.
01 51 36	2.8	CC	Are you going through the day horizon check?
01 51 42	6	P	This is Friendship Seven. Negative, not at present time. I am going to start into a yaw right very shortly. Over.
01 51 52	1.0	CC	Roger, I understand.
01 51 55	3.1	P	This is Friendship Seven. Correction: will do day horizon check now.
01 51 59	1.2	CC	Roger. Standing by.
01 52 09	3.4	P	Friendship Seven, coming up to 14°.
01 52 24	1.5	CC	When are you coming up?
01 52 29	0.4	CC	. . . coming up . . . .
01 52 31	2.5	P	Say again. You're very garbled. This is Friendship Seven.
01 52 34	3.9	CC	Roger. We have you at 3° on the ground on TM readouts.
01 52 39	1.7	P	Roger. Friendship Seven.
01 52 44	1./	CC	We have you about 14 [degrees].
01 52 54	4.2	P	This is Friendship Seven. I have no problem at all controlling on the horizon. Over.
01 53 00	3.7	CC	Roger, understand you have no problem at all controlling with the horizon.
01 53 05	0.8	P	That's affirmative.
01 53 22	1.2	P	This is Friendship Seven.

ATLANTIC SHIP (SECOND ORBIT)

01 53 26	0.8	CC	Friendship Seven, I read you.
01 53 30	3.5	CC	Friendship Seven, Friendship Seven, this is ATS Cap Com. Over.
01 53 34	2.1	P	Hello, ATS Cap Com. Loud and clear; how me?
01 53 38	3.2	CC	Roger, Friendship Seven. You are loud and clear, also.
01 53 48	5.5	P	This is Friendship Seven. My troubles in yaw appear to have largely reversed.
01 53 54	1.7	CC	Friendship Seven, we just lost . . . .
01 53 56	2.6	CC	Roger. Troubles appear to have reversed. Over.
01 53 59	12.7	P	That is affirmative. At one time, I had no left low thrust in yaw, now that one is working, and I now have no low right thrust in yaw. Over.
01 54 14	4.9	CC	Roger, no right yaw, low thrust. Over.
01 54 20	5.1	P	Roger. This is Friendship Seven. Starting 180° yaw right. Over.
01 54 26	4.5	CC	Roger. Starting 180° yaw right. Standing by, go ahead.
01 55 07	0.8	CC	MARK 50°.
01 55 10	1.7	P	This is Friendship Seven. Say again.
01 55 13	1.8	CC	My mark was on 50 degrees.
01 55 15	0.6	P	Roger.
01 55 27	0.8	CC	MARK, a hundred.
01 55 29	0.5	P	Roger.
01 55 52	6.0	P	This is Friendship Seven. Have yawed 165 [degrees], holding orbit attitude in roll and pitch. Over.
01 55 59	4.3	CC	Roger, I confirm, confirm your values. Over.
01 56 05	1.2	P	Roger. Friendship Seven.
01 56 12	3.2	P	This is Friendship Seven, holding 180° in yaw. Over.
01 56 17	3.3	CC	Roger, holding 180° in yaw.
01 56 27	1.6	P	This is Friendship Seven.
01 56 33	1.0	CC	Go ahead, Friendship Seven.
01 56 35	5.2	P	This is Friendship Seven. I like this attitude very much, so you can see where you're going. Over.
01 56 41	3.4	CC	Roger, say you liked your attitude? Over.
01 56 45	1.4	P	Say again, please. Over.
01 56 49	6.0	CC	Friendship Seven, this is ATS Cap Com, suggest you . . . . Over.

01 57 01	4.2	P	This is Friendship Seven. I have a loose bolt floating around inside the periscope.
01 57 13	& 6	CC	Did not read, did not read. Say again, say again.
01 57 25	3.5	P	This is Friendship Seven, yawing back to orbit attitude. Over.
01 57 29	3.3	CC	Roger, coming back to orbit attitude. Over.

KANO (SECOND ORBIT)

01 57 38	2.7	CC	Friendship Seven, this is Kano Cap Com, standing by.
01 57 42	2.3	P	Hello, Kano. Roger. Hear you loud and clear.
01 57 50	6.3	CC	Friendship Seven, this is Kano. Any time you're ready, we'd like your status report. We hear you loud and clear, also.
01 57 59	9.6	P	Roger, Kano. This is Friendship Seven. I am presently returning to orbit attitude from 180° yaw check. Over.
01 58 08	9.5	CC	Roger, Friendship Seven. We marked that your part of the transmission pretty clearly over ATS, and understand you like the old attitude.
01 58 18	4.5	P	Negative, I like the forward-facing attitude much better. Like . . . . .
01 58 24	5.0	CC	But, how is your status, and your station report?
01 58 29	12.2	P	This is Friendship Seven. My status is good. Control fuel is 76-100 [percent]; oxygen, 68-100 [percent]; amps, 25. Over.
01 58 44	10.8	CC	Roger, Friendship Seven. We are monitoring your inverters. Your 150-volt inverter 200°, your 250, 180°-180°.
01 58 57	1.1	P	Roger. Friendship Seven.
01 59 10	9.8	CC	Friendship Seven, we have telemetry solid and check all your systems out okay. We will remind you to start pre-darkside check list as soon as you lose contact with us.
01 59 20	1.0	P	Roger. Friendship Seven.
01 59 23	2.1	P	This is Friendship Seven, back in orbit attitude. Over.
01 59 26	2.3	CC	Roger, understand back in orbit attitude.
01 59 29	1.3	P	This is Friendship Seven. That's affirm.
01 59 32	8.9	P	This is Friendship Seven. I now have control in low thrust to the left. I do not have control in low right thrust. Over.
01 59 42	6.5	CC	Roger. Confirm you have control in low thrust to the left, but no control in low thrust to right.
01 59 49	1.6	P	That is affirmative. Friendship Seven.
01 59 51	3.9	CC	We will inform MCC. Have you any other comments or queries?
01 59 55	5.4	P	Negative. Friendship Seven. I'm in orbit attitude. I'll try and go back on ASCS and see if it works. Over.
02 00 02	1.3	CC	Roger. Will stand by.
02 00 03	0.4	P	Roger.
02 00 28	2.6	p	This is Friendship Seven. The capsule appears
02 00 38	2.5	CC	Friendship Seven, your last transmission was interrupted.
02 00 41	7.9	P	This is Friendship Seven. The capsule appears to be holding in ASCS okay, but I believe we are off in yaw. Do you confirm? Over.
02 00 50	0.6	CC	Standby.
02 00 53	3.7	CC	Our signal shows you are sitting at zero in yaw.
02 01 03	16.2	P	This is Friendship Seven. I'm going to depart from flight plan for a moment and try and work this out a little better here, we are drifting in yaw. I am cutting automatic yaw off. And, and will control manually in yaw temporarily. Over.
02 01 20	5.0	CC	Roger, we still confirm you are holding in yaw, at, as last TM check.
02 02 28	4.6	CC	Friendship Seven, this is Kano Cap Com. I do not think I can read you, handing over to Zanzibar.
02 02 37	1.5	P	Roger, this is Friendship Seven.

ZANZIBAR (SECOND ORBIT)

02 02 49	6.1	CT	Friendship Seven, Friendship Seven this is Zanzibar Com Tech, transmitting on HF UHF. Do you read? Over.
02 02 56	1.8	p	Roger, Zanzibar. Friendship Seven.
02 03 10	6.3	CT	Friendship Seven, Friendship Seven, this is Zanzibar Com Tech, transmitting on HF UHF. Do you read? Over.
02 03 16	3.3	p	Roger, Zanzibar. Read you loud and clear; how me? Over.
02 03 30	2.8	P	Zanzibar Com Tech, this is Friendship Seven. Over.
02 03 40	1.0	P	Zanzibar Com Tech.
02 03 42	5.2	CT	Friendship Seven, this is Zanzibar Com Tech transmitting on HF UHF. Do you read? Over.
02 03 47	3.6	p	Roger, Zanzibar. Friendship Seven. Hear you loud and clear; how me?
02 04 08	3.7	p	Zanzibar, this is Friendship Seven. Do you receive? Over.
02 04 14	8.8	CC	Friendship Seven, this is Zanzibar Cap Com, reading you weak but readable. We have solid telemetry contact. Report your status. Over.
02 04 22	37.2	P	Roger. This is Friendship Seven. My status: I am on ASCS, it is not holding all the time. My trouble in yaw has reversed. During the first part of the flight, when I had trouble over the west coast of the United States, I had a problem with the yaw, with no low thrust to the left; now I have thrust in that direction but do not have low thrust to the right. When the capsule drifts out in that area, it hits high thrust and drops into orientation mode, temporarily. Over.
02 05 01	7.5	CC	Roger, Friendship Seven. This is Zanzibar Cap Com. Continue with your 30 minute report. Over.
02 05 13	1:30. 2	P	Roger. This is Friendship Seven. Thirty-minute report follows: Head movements cause no nausea or no bad feelings at all. I am surprised that I can look as close to the sun as I can; the sun is shining directly on my face at present time, and all I have to do is to shade my eyes with my eyebrow. All fuse switches are still as previously reported on left panel. ASCS Mode Select is normal, auto, gyro normal; Squib is armed, Auto Retrojettison is off. All "T" handles are in except manual and I have it pulled so that I can immediately go to manual as a backup in case the ASCS malfunctions further. The Rescue Aids are normal. Landing bag is off. All other sequence panel items are in normal position. Control fuel is 68-100 [percent], repeat, 68-100. EPI is indicating my approximate position. Time from launch is 02 plus 06 plus 30 MARK. Retrograde time for area H is still set for 04 plus 32 plus 28. I would like to know if Canaveral wants me to reset that into the clock. Over.
02 06 45	2.2	p	To reset the correct time on the clock. Over.
02 06 50	1.4	CC	Standby one, Friendship Seven.
02 06 53	0.5	p	Roger.
02 07 03	7.9	CC	Friendship Seven, this is Zanzibar Cap Com. We don't have word from Canaveral instructing us to instruct you to change the clock at this time.
02 07 11	2.1	P	Roger, Roger.
02 07 20	2.7	CC	Friendship Seven, continue with 30-minute report.
02 07 24	54.2	P	Roger. This is Friendship Seven. Thirty-minute report continues. Cabin pressure is 5.6 and holding. Cabin temperature is 95 [degrees] and dropping slowly. The relative humidity is 20 [percent], coolant quantity is 67 [percent], and suit environment temperature is 68 [degrees], pressure is 5.8. Steam temperature is 47 [degrees], and I'm turning that down a little bit more now. Oxygen is 68-100 percent; amps, 22; voltages main 24, 25, 25, 25; standby one, 25, 25. Cabin excess water light is on, turning that one down. Over.
02 08 20	1.8	CC	Roger, Friendship Seven. Continue.
02 08 23	1.1	P	Roger. Friendship Seven.
02 08 27	3.5	P	Standby one, I need to reorient the capsule. We're getting too far out. Over.

02 08 32	0.4	CC	Roger.
02 08 40	46.3	P	This is Friendship Seven. Sun, I want to make a mark here when the sun goes down. Sun is on the horizon at the present time, a brilliant blue out from each side of it. And I'll give a mark at the last. The sun is going out of sight. Ready now, MARK (2:09:06). There's brilliant blue out on each side of the sun, horizon to horizon almost. I can see a thunderstorm down below me somewhere, and lightning.
02 09 44	8.4	P	And I am not night adapted right now, so I cannot see any of the zodiacal light.
02 09 56	9.0	CC	Friendship Seven, this is Zanzibar Cap Com. Contingency time for Area 2 Delta, 02 38 31. Did you copy? Over.
02 10 07	2.0	P	Say again, contingency area time. Over.
02 10 10	3.3	CC	02 38 31. Over
02 10 14	2.2	P	02 38 31, Roger.
02 10 20	5.7	CC	Friendship Seven, this is Zanzibar Cap Com. Please comment on reach accuracy and visual acuity. Over.
02 10 26	27.4	P	Right, visual acuity is good. I have no problem reading the charts, no problem with the astigmatism at all. I am having no trouble at all holding attitudes either. I'm still on fly-by-wire. I'm on normal ASCS but I'm backing it up with manual at present time. Over.
02 10 56	4.3	CC	Roger, Friendship Seven. We have intermittent pitch ignore.
02 11 02	11.7	CC	The inverter temperatures are 205° for fans inverted 190° for ASCS inverter. We are losing telemetry contact. Stand by to contact IOS, Zan.
02 11 15	1.1	P	Roger. Friendship Seven.
02 11 33	5.1	P	This is Friendship Seven, Zanzibar. Does Cape recommend any action on inverters? Over.
02 12 04	3.9	P	This is Friendship Seven. Have a lot of lightning under me at the present time.
02 12 50	9.1	P	This is Friendship Seven. I still have very bright light along the horizon; orange at the bottom, yellowish layer, then blue, then very dark on top.

INDIAN OCEAN SHIP (SECOND ORBIT)

02 13 18	5.7	CT	Friendship Seven, Friendship Seven, this is Indian Com Tech on HF and UHF How read? Over.
02 13 24	18.9	P	This is Friendship Seven. I read you okay, Indian Com Tech. This is Friendship Seven. I am on straight manual control at present time. I have 92 percent fuel. Over on, on manual, auto fuel is at 64 [percent]. Over.
02 13 57	8.1	CC	Friendship Seven, Friendship Seven, this is Indian Cap Com. You are fading in and out. Say again your fuel positions. Over.
02 14 00	3.1	P	This is Friendship Seven. Standby one.
02 14 04	0.8	CC	Roger standing by.
02 14 10	9.9	P	This is Friendship Seven. Capsule dropped into orientation mode again on ASCS. I took it over manually and am reorienting at present time. Over.
02 14 21	2.7	CC	Roger, understand you're reorienting in manual. Over.
02 14 24	15.5	P	Right, it dropped into orientation mode. I have fuel quantity warning light on in automatic. I'm okay at present time. Have capsule under control and using manual. Over.
02 14 36	3.3	CC	Roger, understand. Will you give me a list of your consumables, please? Over.
02 14 40	1.1	P	Roger. Standby one.
02 14 45	15.7	P	This is Friendship Seven. Fuel, 62-90 [percent]; oxygen, 68-100 [percent]; amps, 20. Over.
02 15 02	3.0	CC	Roger, Roger. You're coming in loud and clear, loud and clear. Over.
02 15 06	8.5	P	Roger. This is Friendship Seven. I have good night horizon check, can control okay on night horizon.

02 15 18	0.9	CC	Roger, understand.
02 15 28	4.4	CC	Friendship Seven, this is Indian Cap Com. Can you see any constellations and identify them? Over.
02 15 33	9.9	P	This is Friendship Seven. Affirmative. I'm pitching up at present time to try and identify some of them. At capsule attitude in orbit, you can't see very much of the sky. Over.
02 15 44	1.1	CC	Roger, understand.
02 16 09	4.8	CC	Friendship Seven, this is Indian Cap Com. Are you prepared to copy your retro sequence time? Over.
02 16 15	2.1	P	This is Friendship Seven. Standby one.
02 16 18	0.4	CC	Roger.
02 17 19	3.9	CC	Friendship Seven, this is Indian Cap Com. Be advised we have flare ignition. Over.
02 17 24	4.9	P	Roger. This is Friendship Seven. Do not have it in sight; cloud cover solid underneath. Over.
02 17 30	1.1	CC	Roger, understand.
02 18 13	5.3	P	This is Friendship Seven. IOS, do you have any indication of fuel temperatures? Over.
02 18 20	0.8	CC	Standby one.
02 18 45	10.1	CC	Friendship Seven, this is Indian Cap Com. We read 95° on auto fuel temp, 95° on manual fuel temp. Standby one.
02 19 01	0.4	P	Roger.
02 19 03	3.3	CC	We read 65° on manual fuel temp. Over.
02 19 07	3.0	P	Roger, understand 65° on manual fuel temp.
02 19 12	8.9	CC	That is affirmative. We have message from MCC for you to keep your Landing Bag switch in off position, Landing Bag switch in off position. Over.
02 19 21	1.4	P	Roger. This is Friendship Seven.
02 19 25	3.1	CC	Roger. Are you prepared to copy retrosequence times? Over.
02 19 28	0.6	P	Standby one.
02 19 31	1.5	P	Okay. Friendship Seven. . . .
02 19 35	4.1	CC	Roger. Do you have Area 2 Delta time, Area 2 Delta? Over.
02 19 39	1.3	P	Friendship Seven. Negative.
02 19 41	8.8	CC	Roger. Area 2 Delta time is 2 hours, 38 minutes, 31 seconds. I say again, 2 hours, 38 minutes, 31 seconds. Over .
02 19 50	2.6	P	Roger. Friendship Seven. 2 plus 38 plus 31.
02 19 54	5.9	CC	Roger. Area 2 Echo is 2 hours plus 48 minutes plus 69 seconds.
02 20 00	0.5	P	Say,
02 20 02	2.4	CC	48 minutes, plus 59 seconds. Over.
02 20 05	3.3	P	2 hours plus 48 minutes plus 59 seconds for 2 Echo.
02 20 10	11.1	CC	Roger. End of orbit area Golf is 3 hours plus 00 minutes plus 39 seconds, I say again, 3 hours plus 00 minutes plus 39 seconds. Over.
02 20 23	4.2	P	Roger, 03 plus 00 plus 39 seconds.
02 20 27	1.2	CC	Roger, that is affirmative.
02 20 37	4.2	CC	Friendship Seven, this is Indian Cap Com. Have you noticed any constellations yet? Over.
02 20 41	12.7	P	This is Friendship Seven. Negative. I have some problems here with ASCS. My attitudes are not matching what I see out the window. I've been paying pretty close attention to that; I've not been identifying stars. Over.
02 20 55	0.9	CC	Roger, understand.

MUCHEA (SECOND ORBIT)

02 21 19	6.8	CT	Friendship Seven, Friendship Seven, this is Muchea, Com Tech. Friendship

			Seven, this is Muchea, Com Tech. Do you read?
02 21 28	2.5	P	Hello, Muchea Com Tech, Friendship Seven. Over.
02 21 45	6.9	CT	Friendship Seven, Friendship Seven, this is Muchea Com Tech. Have your telemetry. Muchea Com Tech. Do you read?
02 21 51	4.0	P	Hello, Muchea Com Tech. Friendship Seven. Read you loud and clear; how me?
02 22 14	5.9	CC	Friendship Seven, Friendship Seven, this is Muchea Com Tech. Friendship Seven, this is Muchea Com Tech. Do you read? Over.
02 22 24	5.7	P	Hello, Muchea Com Tech, Friendship Seven. I read you loud and clear; how me?
02 22 38	5.3	P	Muchea Com Tech, Friendship Seven. Holding attitude on Orion. Over.
02 22 46	7.5	CC	Friendship Seven, Friendship Seven, this is Muchea Com Tech, calling on HF and UHF. Friendship Seven, this is Muchea Com Tech. Do you read?
02 23 10	3.3	P	Hello Muchea, hello Muchea. Friendship Seven on HF. Over.
02 23 16	2.6	CT	Friendship Seven, this is Muchea Com Tech. Do you read?
02 23 19	3.2	P	Roger, Muchea. Friendship Seven. Read you loud and clear; how me?
02 23 24	2.7	CC	Friendship Seven, would you say again?
02 23 29	3.2	P	This is Friendship Seven, Muchea. I read you loud and clear. Over.
02 23 34	3.3	CC	Roger, Friendship Seven. I'm reading you loud and clear on HF. Over.
02 23 38	0.6	P	Roger.
02 23 42	2.7	CC	Friendship Seven, Muchea Cap Com. How me? Over.
02 23 45	12.3	P	Roger, Muchea Cap Com. Loud and clear. Fuel, 62-85 [percent]; oxygen, 65-100 [percent]; amps, 22. Over.
02 24 00	3.2	CC	Roger. I did not get your fuel. Would you give that ... to us again? Over.
02 24 03	3.8	P	Roger, fuel is 60-85 [percent]. Over.
02 24 24	2.6	CC	Friendship Seven, how do you read us on UHF? Over.
02 24 28	2.9	P	Roger. Loud and clear; I am still on HF. Over.
02 24 33	4.1	CC	Friendship 7, Muchea Cap Com. Recommend you go UHF. Over.
02 24 37	1.4	P	Roger, going to UHF.
02 24 41	2.3	CC	Friendship Seven, Muchea Cap Com. How now? Over.
02 24 49	2.9	CC	Friendship Seven, this is Muchea Cap Com. How are read now? Over.
02 24 52	3.4	P	Hello, Muchea. Friendship Seven. Read you loud and clear; how me?
02 24 56	4.4	CC	Loud and clear. You were coming in slightly garbled. Could you give us your fuel, oxygen and amps? Over.
02 25 01	6.4	P	Roger. Fuel, 60-85 [percent]; oxygen, 65-100 [percent]; amps, 22. Over.
02 25 09	2.5	CC	Roger. Exhaust temperature?
02 25 12	2.1	P	Exhaust temperature 50 [degrees]. Over.
02 25 15	1.9	CC	Roger. Are you in fly-by-wire?
02 25 17	10.4	P	Negative, I am on manual at present time. I'm down to 60 [percent] on automatic fuel, so I cut it off and I'm on manual at present time. Over.
02 25 28	2.6	CC	Roger. How is your yaw thruster problem? Over.
02 25 32	19.4	P	This is Friendship Seven. I am getting some erratic indications in all axes. When I align everything on orbit attitude by the instruments I am considerably off where I should be. I'm rolled some 20° to the right; I'm also yawed to the right a little bit. Over.
02 25 52	6.3	CC	Roger, understand. Are you satisfied with the fly-by-wire and manual proportional systems? Over.
02 25 58	4.2	P	Fly-by-wire is not functioning in yaw low right. Over.
02 26 06	1.9	CC	Roger, low yaw right.
02 26 08	13.6	P	That's affirmative. When I first had trouble with it, it was malfunctioning in low yaw left, and now low yaw left is operating okay, but low yaw right is not operating. Over.
02 26 22	5.1	CC	Roger. Have you tried caging your gyros and re-erecting them. What do you think about this? Over.
02 26 27	12.1	P	Negative, have not yet. I want to get an alignment as soon as I get back in

daylight and do just exactly that. I think my, I'm probably not—my scanners are probably not working right. Over.

02 26 40	5.1	CC	Roger. Will you confirm the Landing Bag switch is in the off position? Over.
02 26 45	3.6	P	That is affirmative. Landing Bag switch is in the center off position.
02 26 50	4.3	CC	You haven't had any banging noises or anything of this type at higher rates?
02 26 55	0.4	P	Negative.
02 26 57	6.4	CC	They wanted this answer. Do you have your retrosequence times 2 Dog, Easy, and Golf from Indian Ocean Ship?
02 27 03	6.6	P	Standby one. Yes, 2 Easy, and Golf, yes, I do have those.
02 27 11	.4	CC	Roger.
02 27 14	3.1	CC	Do you have any, are you ready for Z and R Cal? Over.
02 27 18	1.2	P	Affirmative, go ahead.
02 27 20	1.1	CC	Cal coming through.
02 27 26	2.5	CC	Any comments on weightlessness, etc.
02 27 30	5.7	P	Negative, I have no ill effects whatsoever. I don't even really notice it now. I'm just, very comfortable.
02 27 37	0.3	CC	Fine.
02 27 38	4.9	P	I have the constellation of Orion up here in the window now. I'm holding a position on it.
02 27 45	6.3	CC	Roger, good. Your 150 VA is 200 [degrees]; your 250 VA is 190 [degrees]. Over.
02 27 51	0.4	P	Roger.
02 27 52	1.2	CC	R Cal, coming now.
02 28 11	5.2	CC	Friendship Seven, Muchea, Cap Com. Do you still have high, both high thrusters in fly-by-wire? Over.
02 28 17	1.2	P	That's affirmative, I do.
02 28 19	0.6	CC	Roger.

WOOMERA (SECOND ORBIT)

02 28 35	6.6	CC	Friendship Seven, Woomera Cap Com. We have contact, TM solid and UHF solid. Over.
02 28 40	0.7	P	Roger, Woomera.
02 28 45	2.2	CC	Friendship Seven, are you ready to go ahead with this 30-minute report? Over.
02 28 49	2.1	P	Negative, not just at the moment. Stand by.
02 28 52	0.9	CC	Roger, Friendship Seven.
02 28 58	5.3	CC	Friendship Seven, Muchea. Just a reminder to take your exercise after the blood pressure. Over, Woomera.
02 29 04	0.6	P	Roger.
02 29 46	2.0	CC	Friendship Seven, Woomera Cap Com. Do you read? Over.
02 29 48	0.8	P	Go ahead, Woomera.
02 29 50	5.8	CC	The Cape advises that you conduct this horizon check and cage the gyros, before California.
02 29 58	3.5	P	Say again, please, Woomera, did not understand.
02 30 01	7.3	CC	Cape advises that you conduct this horizon check, and cage the gyros before California contact. Over.
02 30 09	2.4	P	Roger, understand before California contact.
02 30 13	6.8	CC	Roger. Would like to confirm your control systems status, fly-by-wire low yaw right is inoperative.
02 30 20	0.9	P	That's affirmative.
02 30 22	1.3	CC	Good in all axes. Over.
02 30 24	0.9	P	That is affirmative.
02 30 27	2.4	CC	Can you give status on your ASCS system? Over.
02 30 30	52.6	P	This is Friendship Seven. ASCS started out operating, it would drop into

orientation mode in yaw and that would throw the other axes out. It was doing this for a while it would drift off slowly, yaw right and there would be no correction until it went out about 20°. Then it would come rapidly back across and correct itself from a 20° right attitude, would come back across to a left attitude in orientation mode. This continued for a while. I took control, that was right after Guaymas. By the time I had gone over Africa the condition had just reversed, had just reversed. The same thing, only the other direction. Over.

02 31 24	0.9	CC	Roger, I understand.
02 31 28	3.5	CC	Friendship Seven, if you're ready, could we conduct this 30 minute report? Over.
02 31 32	27.2	P	Friendship Seven. Roger. All switch fuses still same position. Squib is armed. Auto Retrojettison is off. ASCS mode selector is normal, rate command, gyro, normal. Manual fuel handle is pulled. Automatic fuel is still on. Retro-Delay, normal, Landing Bag is off. Everything else on sequence panel is normal. Fuel . . .
02 32 00	7.1	CC	Could we have a blood pressure check? Start your button, then exercise the bungee cord, then start your blood pressure check again. Over.
02 32 08	2.7	P	Roger. Friendship Seven. Blood pressure coming on.
02 32 26	5.9	CC	Roger, Friendship Seven. We're reading your blood pressure, what kind of result on your visual acuity test? Over.
02 32 32	6.7	P	Roger. Friendship Seven. Visual acuity still good. Can read same lines I did when starting out. Over.
02 32 40	6.8	CC	Roger. Would like to go through this 30-minute report again, starting with the Auto Retro-jett switch. Go ahead.
02 32 49	7.7	P	Auto Retro-jettison is armed. I'm having to break this in the middle because I need to keep looking out the window to control accurately. Over.
02 32 57	2.7	CC	Roger, Friendship Seven. discussion. Over.
02 32 00	0.5	P	Roger.
02 33 09	57.8	P	This is Friendship Seven. Auto Retro-jettison is off. ASCS mode selector is normal, rate command, gyro normal. All "T" handles are in except manual, and it is out. I am controlling manually at present time. Landing Bag switch is off. Sequence panel is all normal except for that. The control fuel is 60-82 [percent]. Attitudes at present time on manual control are roll 5 [degrees] right, yaw 15 [degrees] right, pitch -34 [degrees] in orbit attitude. My time from launch is 2 plus 33 plus 60, MARK! Correction that would be 2 plus 34 on that mark. Did you receive? Over.
02 34 09	0.2	CC	Yes.
02 34 10	9.0	P	Roger. Cabin pressure is 5.5 and holding. Cabin air is 95 [degrees]. Cabin excess water light is still on. I'm turning it down a little bit more yet.
02 34 34	50.3	P	Cabin relative humidity is indicating 20 [percent]. Coolant quantity is 66 [percent]. Suit is 68 [degrees] on temperature, pressure 5.8 indicated, steam temperature, 50 [degrees] in suit circuit and comfortable. Oxygen is 65-100 [percent] amps 20. Voltages: main 24, 25, 25, 25; standby one is 25, 25; isolated 29; and back on main. Fans is 113, ASCS 113. I have two warning lights on, the excess cabin water and fuel quantity. Over.
02 35 29	6.2	CC	Roger, Friendship Seven. Aeromed would like to know whether you have conducted your exercise on bungee cord. Over.
02 35 36	6.1	P	Negative, I have not done that recently. Will conduct that now as part of this check. You had a blood pressure.
02 35 43	5.3	CC	You are fading now. Standby for Canton in a few minutes. This is Woomera Cap Com, Over and Out.
02 36 02	2.9	P	Okay, exercise conducted. Blood pressure starting.
02 36 36	4.9	P	This is Friendship Seven. Woomera, are you still receiving me? Over.
02 37 07	1.3	P	This is Friendship Seven.

CANTON (SECOND ORBIT)

02 42 01	2.8	CT	Friendship Seven, Canton Com Tech.
02 42 05	4.2	P	Hello Canton Com Tech. Friendship Seven. How did you receive UHF-HI? Over.
02 43 10	5.3	CT	Friendship Seven, Canton Com Tech. Reading you loud and clear. Standby for Cap Com.
02 43 16	24.9	P	Roger. Canton Cap Court, Friendship Seven. Fuel 62-75 [degrees], oxygen 65-98 [percent], correction 65-100 [percent], amps 23. My condition is good. The sunrise is coming around the capsule right now. Over.
02 43 42	5.2	CC	Roger, Friendship Seven. This is Canton. We have TM solid; go ahead, Over.
02 43 48	7.1	P	Roger. This is Friendship Seven, and now that the sunrise is starting, I have all these little particles coming around the capsule again, just at sunrise.
02 43 57	0.9	CC	Roger, Friendship Seven.
02 44 02	5.3	P	I also can see the light on my, on steam from the thruster when I operate it. Over.
02 44 09	7.2	CC	Roger, Friendship Seven. Are, are your thrusters working, are all your high thrusters working okay. Over.
02 44 17	4.1	P	This is Friendship Seven. Affirmative, operating okay.
02 44 27	9.8	P	This is Friendship Seven. I think my, I can see a little bit of steam spitting against the dark sky here occasionally from my pitch down manual thruster. Over.
02 44 36	0.4	CC	Roger.
02 44 37	12.7	P	This is Friendship Seven. All these little particles; there are thousands of them; and they're not coming from the capsule. They're something that's already up there because they're all over the sky. Way out I can see them, as far as I can see in each direction, almost.
02 44 52	1.1	CC	Roger, Friendship Seven.
02 44 56	5.5	CC	Friendship Seven, this is Canton. Our telemetry here indicates that everything is okay. Over.
02 45 02	1.2	P	Roger, Friendship Seven.
02 45 05	8.9	CC	Friendship Seven, this is Canton. Our Aeromed would like to hear any comments you have on weightlessness, nausea, dizziness, taste and smell sensations. Over.
02 45 16	32.3	P	Friendship Seven, I have no sensations at all from weightlessness except very pleasant. No ill effects at all, I'm not sick; feel fine. I have had no, no dizziness; I've run the head checks, the head movements and have no problem with that. I've run the oculogyric check and have no problem with that. I haven't been able to use much of the equipment this orbit, however. I've been mainly concerned with this control problem. Over.
02 45 48	11.6	CC	Roger, Friendship Seven. Our fuel quantities here agree with yours. Our 150 fans inverter temperature is 200 [degrees]. The ASCS temperature is 190 [degrees]. Out
02 46 00	1.2	P	Roger, Friendship Seven.
02 46 14	4.6	CC	Friendship Seven, this is Canton. Do you have all the retro times that you need? Over.
02 46 19	10.3	P	This is Friendship Seven. I believe so. I have Foxtrot coming up here, I mean Golf coming up, at 03 00 39. Is that affirm?
02 46 30	0.6	CC	Standby one.
02 46 35	4.8	P	This is Friendship Seven. I'm trying to get some pictures of these particles that are outside here. Over.
02 46 50	5.8	CC	Friendship Seven, that time we have for Golf is 03 00 39. Over.
02 46 58	0.4	P	Roger.
02 47 14	6.1	CC	Friendship Seven, this is Canton. We also have no indication that your

			landing bag might be deployed. Over.
02 47 22	3.1	P	Roger. Did someone report landing bag could be down? Over.
02 47 26	8.7	CC	Negative, we had a request to monitor this and to ask you if you heard any flapping, when you had high capsule rates ....
02 47 36	0.5	P	Negative.
02 47 37	1.1	CC	... this was. Over.
02 47 38	16.5	P	Well, I think they probably thought these particles I saw might have come from that, but these are, there, there are thousands of these things, and they go out for, it looks like miles in each direction from me and they move by here very slowly. I saw them at the same spot on the first orbit. Over.
02 47 56	0.4	CC	Roger.
02 48 01	2.4	CC	Friendship Seven, what control mode are you in presently?
02 48 04	11.9	P	This is Friendship Seven. I am in manual. While on the dark side, I aligned myself with the horizon, when up to 0 0 0, and caged the gyros and uncaged again. Over.
02 48 17	9.9	CC	Roger, Friendship Seven. Will be losing contact any minute now, you have reentry checks to make and approximately 10 minutes, until California contact. Over.
02 48 27	1.0	P	Roger, Friendship Seven.

HAWAII (SECOND ORBIT)

02 48 36	4.2	CT	Friendship Seven, this is Hawaii Com Tech on UHF and HF. How do you read? Over.
02 49 42	2.4	P	Say again, Com Tech, what station?
02 49 06	3.8	CT	Friendship Seven, Hawaii Com Tech on UHF and HF. How do you read? Over.
02 49 10	3.1	P	Hello Hawaii. Loud and clear; how me? Over.
02 49 22	3.2	P	Hello Hawaii. Friendship Seven. Loud and clear; how me? Over.
02 49 27	5.9	CC	Friendship Seven, this is Hawaii Cap Com. I read you loud and clear. What is your status report? Over.
02 49 33	1.4	P	Roger, Friendship Seven, ah.
02 49 36	2.2	---	He has contact, we're putting air-to-ground on.
02 49 39	12.7	P	This is Friendship Seven. Fuel 62-74 [percent], oxygen 64-98 [percent], amps 23. Over.
02 49 53	4.5	CC	Roger, Friendship Seven. The first part of your transmission was broken up; could you say again, fuel?
02 49 59	3.5	P	Roger, fuel 62-74 [percent]. Over.
02 50 03	14.0	CC	Roger. Read 62-74 [percent], Friendship Seven. Cape would like to know if you have made the visual check on the gyros yet?
02 50 12	13.8	P	This is affirmative. During the dark side I went to 0 0 0 as best I could and caged and uncaged the gyros. I'm rechecking this at present time. Over.
02 50 26	3.7	CC	Roger. I'll stand by until you complete your checks and we'd like to get the results.
02 50 30	1.0	P	Roger, Friendship Seven.
02 50 52	4.2	P	Hawaii, this is Friendship Seven. Are we getting scanner ignore? Over.
02 50 58	4.7	CC	There are no ignore indications on the ground as of now, Friendship Seven. Hawaii. Over.
02 51 04	8.5	P	Roger. This is Friendship Seven. When I aligned in, standby one.
02 51 53	11.2	P	This is Friendship Seven. Roll appears to be about 20° off. At present time, I am indicating 20° right roll when I am lined up perfectly with the horizon. Over.
02 52 06	7.0	CC	Roger, Friendship Seven. Understand you still have a 20° error in your attitude indication after caging the gyros.
02 52 13	8.8	P	This is Friendship Seven. That's affirmative. It's drifted off again. It was

			correct at that time; it apparently has drifted off. It's about 20° off in roll.
02 52 25	4.1	CC	Friendship Seven, could you give us a readout on all three axes; we'd like to compare ground readouts.
02 52 29	7.6	P	Roger, Friendship Seven. Roll indicates 19 [degrees] right, yaw, 2 [degrees] left, pitch - 22 [degrees]. Over.
02 52 38	1.8	CC	Roger, Friendship Seven. Read you loud and clear.
02 52 50	3.7	P	Friendship Seven, Hawaii Cap Com. Can you estimate the drift rates you are getting?
02 52 54	10.0	P	This is Friendship Seven. Drift rates are very difficult to pick up for some reason. I have not been able to get real accurate drift rates. I feel that . . . .
02 53 06	4.3	CC	Do you still consider yourself GO for the next orbit?
02 53 10	2.4	P	That is affirmative, I am GO for the next orbit.
02 53 13	2.4	CC	Roger, I understand. At present time ground concurs.
02 53 16	0.4	P	Roger.
02 53 24	3.9	CC	Could you, could you give me a readout on exhaust temperature, please?
02 53 28	3.2	P	Friendship Seven. Roger. Exhaust temperature, 47 [degrees].
02 53 36	3.0	CC	Friendship Seven, Hawaii Cap Com. Could you read exhaust temperature please?
02 53 40	1.9	P	Roger. Exhaust temperature 47 [degrees].
02 53 42	1.6	CC	Roger. Understand 47 [degrees].
02 53 46	3.4	CC	Surgeon requests a blood pressure readout if convenient at this time.
02 53 51	0.6	P	Roger 47 [degrees].
02 53 55	6.4	P	Correction, Roger, blood pressure, 47 [degrees] on steam temp, blood pressure on.
02 54 24	9.8	P	This is Friendship Seven. I have a pretty good run here on, on yaw check. I believe my yaw is pretty accurate. Roll is off by about 20° to the right at present time.
02 54 35	5.9	CC	Roger, Friendship Seven. Roll is off by 20° to the right; yaw looks pretty good. Hawaii. Over.
02 54 42	0.7	P	That's affirmative.
02 54 55	5.7	CC	Friendship Seven, Hawaii Cap Com. MCC confirms that they are GO at the present time for third orbit.
02 55 01	0.9	P	Roger. Friendship Seven.
02 55 14	2.1	CC	Friendship Seven, Hawaii Cap Com. Do you read? Over.
02 55 16	2.1	P	Roger, Hawaii. Loud and clear, how me?
02 56 05	1.2	CC	Are you ready to copy?
02 56 08	1.4	P	Friendship Seven. Go ahead.
02 56 31	13.5	P	This is Friendship Seven. I have a better alignment now on the day side, I believe, on these, on the attitude. I will probably cage and uncage gyros again. Over.

CALIFORNIA (SECOND ORBIT)

02 56 58	5.7	CT	Friendship Seven, Friendship Seven, this is California Com Tech, California Com Tech. Do you read? Over.
02 57 05	3.6	P	Roger, California Com Tech. Read you loud and clear. How me? Over.
02 57 16	3.3	P	Hello, California Com, Tech. Friendship Seven. How me? Over.
02 57 23	5.9	CT	Friendship Seven, Friendship Seven, this is California Com Tech, California Com Tech. Do you read? Over.
02 57 30	3.3	P	Roger, California Com Tech. Friendship Seven. How me? Over.
02 57 48	5.7	CT	Friendship Seven, Friendship Seven, this is California Com Tech, California Com Tech. Do you read? Over.
02 57 54	3.6	P	Hello, California Com Tech. Friendship Seven. Loud and clear; how me?
02 58 14	6.2	CT	Friendship Seven, Friendship Seven, this is California Com Tech, California Com Tech. Do you read? Over.

02 58 21	3.5	P	Hello, California Com Tech. Friendship Seven. Loud and clear; how me?
02 58 25	3.6	CC	Friendship Seven, California Cap Com. Read you loud and clear, John.
02 58 32	3.5	P	Say again. You came in garbled that time. This is Friendship Seven.
02 58 36	4.8	CC	Friendship Seven, this is California Cap Com. We read you loud and clear; how me?
02 58 41	36.4	P	Roger. Receiving you much better now, Wally. Very good. Fuel is 62-62 [percent], oxygen 62-95 [percent], amps are 25. All systems are still go. I have had some erratic ASCS operation. I caged and uncaged on the night side and it appears to be working fairly well now although I was drifting again in roll a moment ago. It appears to have corrected itself in roll, however, without me caging again now. Over.
02 59 17	11.8	CC	Good, John. We have a go all the way on this. I'd like to give you your inverter temperatures. Your fans are 215 [degrees], your ASCS 198 [degrees]. We're not going to do anything about them. Looks real good.
02 59 23	0.6	P	Okay, fine.
02 59 26	4.3	CC	Will you give me your attitudes and we'll check those with ground.
02 59 31	10.6	P	Roger. I'll go back into orbit attitude, drifting toward it at present time. Do you have TM solid now? Over.
02 59 42	0.7	CC	Good here.
02 59 43	6.2	P	Roger. Roll is 5 [degrees] left, yaw 3 [degrees] right, pitch -32 [degrees], right now.
02 59 51	1.0	CC	Roger, have your readings.
03 00 10	4.3	CC	John, we check almost right on the button with your attitudes within 2°.
03 00 14	5.9	P	Roger. I appear to have a little bit of drift in the scope yet though:
03 00 19	2.3	CC	Roger. You probably don't have a good reference yet, do you?
03 00 22	9.8	P	It's rather, it's more difficult to pick up drift than I thought it would be. Your best drift really is to look out the window and try and get something moving away from you out the window.
03 00 36	0.9	CC	Roger, I've got you there.
03 00 43	5.4	CC	You understand that your capsule elapsed time is running about a second slow compared to GET.
03 00 51	4.0	P	No, I did not; I was not aware of that on the elapsed time.
03 00 56	6.3	CC	Roger. This will affect your sequence time. I'll give you 3 Alpha if you're ready to copy.
03 01 03	1.0	P	Standby.
03 01 11	1.7	P	Roger, ready to copy 3 Alpha.
03 01 13	7.8	CC	Roger. 3 Alpha corrected for the second error will be 03 plus 11 plus 26. Over.
03 01 22	2.5	P	Roger, 03 11 26.
03 01 35	4.8	CC	You have your end-of-orbit time, and that has not changed and I believe you have Area Golf also.
03 01 41	9.4	P	That's affirmative. I have Golf at 03 plus 00 plus 39 and Hotel at 04 plus 32 plus 40.
03 01 51	7.2	CC	That is correct. You can correct your Area Golf by subtracting one second and subtract one second from Hotel.
03 01 59	6.4	P	Roger. Understand for my retrosequence for end-of-mission is 04 plus 32 plus 39. Is that affirm?
03 02 06		CC	Negative. That should be 04 plus 32 plus 38.
03 02 10	6.3	P	Plus 38. Roger. Does Cape want me to set that into clock at present time? Over.
03 02 17	3.1	CC	Negative, we have not had those instructions. I'll have Arnie check for you.
03 02 20	5.7	P	Roger. I am still operating on prelaunch time of 04 plus 32 plus 28. Over.
03 02 27	0.3	CC	Understand.
03 02 33	4.8	P	This is Friendship Seven. Coming across the Gulf of Lower California at present time in the scope.

03 02 40	0.3	CC	Roger.
03 02 44	3.7	CC	John, for your information, the clocks will be set over Canaveral.
03 02 49	6.2	P	Roger, understand clocks will be set by command, is that affirm, or do they want. . .
03 02 56	5.4	CC	That is negative. They, on contact with Al, they will have you change the clock.
03 03 01	1.7	P	Roger, understand. Friendship Seven.
03 03 04	6.3	CC	For your information your O2 is holding. It seems to be working; it's reading about 5.2 to 5.3.
03 03 11	12.7	P	Roger, Friendship Seven. Cabin temperature is holding steady at about 90 [degrees]. I still have excess cabin water light on and my, I have the cabin water turned almost completely off at present time.
03 03 25	3.4	CC	Roger, it sounds like it's working out all right anyway.
03 03 28	3.5	P	Roger, it's doing okay. We'll make one more orbit okay.
03 03 32	3.3	CC	I assume you are still on manual proportional; is that correct?
03 03 35	1.7	P	That's affirmative, I'm still on manual.
09 03 40	3.0	CC	Friendship Seven, Guaymas Cap Com reads you loud and clear. Over.
03 03 44	2.2	P	Roger, Guaymas. Read you loud and clear, also.
03 03 54	2.9	CC	John, the Aeromeds are real happy with you; you look real good up there.
03 03 57	3.7	P	All right, fine, glad everything is working out. I feel real good, Wally. No problems at all.

TEXAS (SECOND ORBIT)

03 04 18	3.7	CT	Friendship Seven, Friendship Seven, this is Texas Com Tech. Do you read? Over.
03 04 22	4.3	P	Hello, Texas, this is Friendship Seven. Read you loud and clear. How me? Over.
03 04 27	4.4	CT	Roger, reading you five square on HF. Let's try UHF. Over.
03 04 33	2.4	P	This is Friendship Seven, I'm on UHF. Over.
03 04 37	2.9	CT	Roger, Friendship Seven, Roger. Standby for Cap Com.
03 04 40	0.5	P	Roger.
03 04 51	2.8	CC	This is Texas Cap Com, Friendship Seven. Do you read?
03 04 54	2.7	P	Roger, Texas Cap Com. Loud and clear. How me?
03 04 57	3.9	CC	Loud and clear. We have no queries down here. Continue with your observations please.
03 05 02	12.6	p	Roger. This is Friendship Seven. Systems all operating normally. I gave my quantities again. There's no need to waste more time on those. I am pitching down to about 60° to make observation. Over.
03 05 15	3.1	CC	Roger, we confirm down here on the TM. Over.
03 05 20	0.6	P	Roger.
03 05 54	5.6	P	This is Friendship Seven. Just passed over El Paso in good shape. Could see town through some of the clouds, Over.
03 06 14	3.5	P	Friendship Seven, flying over complete cloud deck at present time.
03 06 24	6.9	P	I'm pitching down to 60° at present time. Orbit attitude in roll and yaw.
03 07 01	0.6	P	This is . . . .

CANAVERAL (THIRD ORBIT)

03 07 03	2.8	CT	Canaveral Com Tech. How do you copy? Over.
03 07 05	3.3	P	Hello, Canaveral Com Tech. Friendship Seven. Loud and clear; how me?
03 07 09	5.1	CT	Roger, Friendship Seven. This is Canaveral Com Tech. Copy you loud and clear, also. Standby for Cap Com please.
03 07 14	0.5	p	Roger.
03 07 16	9.5	P	Hello, Cap Com. Friendship Seven. Fuel is 62-60 [percent]; oxygen is 62-95

[percent]; amps 23. Over.

03 07 28	11.1	CC	Friendship Seven, reading you loud and clear. I'll give you the 3 Bravo Time, 03 22 26. Over.
03 07 42	4.4	P	Roger, 03 22 26, for 3 Bravo. Is that affirm?
03 07 47	7.0	CC	...3 Charlie 03 40 18. Over.
03 07 54	2.0	P	Roger, 03 40 18 for 3 Charlie.
03 08 00	11.2	CC	That is correct. At this time change your retro setting manually to 04 32 38. Over.
03 08 11	3.6	P	Roger, 04 plus 32 plus 38.
03 08 25	4.8	P	Roger, retrograde time is reset to 04 plus 32 plus 38. Over.
03 08 32	25.5	CC	Roger, Seven. We recommend for the third orbit that you use gyros as you desire either normal or free so that in the event prior to retrofire on the third orbit that the scanners and ASCS do not program properly you may use your gyros in the free position for attitude reference. Over.
03 08 58	22.5	P	Roger. This is Friendship Seven. I have a fair, pretty good line-up now on the gyros, I believe. The check that I made on the night side was okay but they drifted off again, apparently rather rapidly in fact. I got another check on it and they seem to have corrected back pretty good now. I did not have to cage them again. Over.
03 09 22	7.5	CC	Roger, Seven, we understand. The only problem is that you may not have enough light time prior to retrofire.
03 09 30	0.6	P	Roger.
03 09 31	5.1	CC	Let the gyros work in the free position if you desire. Over.
03 09 37	1.0	P	Roger. Friendship Seven.
03 09 39	7.9	CC	Also, Seven, we recommend that you allow the capsule to drift on manual control in order to conserve fuel. Over.
03 09 48	1.2	P	Roger. Friendship Seven.
03 09 55	2.4	CC	Seven, this is Cape. Standby for Z Cal.
03 09 58	0.5	P	Roger.
03 10 08	0.2	CC	R Cal.
03 10 20	0.3	CC	Cal off.
03 10 21	0.6	P	Roger.

BERMUDA (THIRD ORBIT)

03 10 30	2.6	CC	Friendship Seven, this is Bermuda Cap Com. We read you loud and clear.
03 10 33	2.3	P	Roger, Bermuda. Hear you loud and clear also.
03 10 36	1.3	CC	Seven, this is Cape. Over.
03 10 38	0.7	P	Go ahead, Cape.
03 10 40	6.7	CC	Correct your 3 Bravo time to 03 22 22. Over.
03 10 48	4.0	P	Roger, 03 22 22 for 3 Bravo.
03 10 51	3.8	CC	Well, Seven, I'm having trouble, the seconds should be 32. Over.
03 10 57	2.7	P	Roger, 03 22 32.
03 11 02	0.5	CC	Good Show.
03 11 04	2.9	P	Roger. Roger.
03 11 11	2.2	P	Hello Bermuda. Friendship Seven. It's over to you. Over.
03 11 14	3.4	CC	Roger, Friendship Seven. We have nothing for you, you're in good shape.
03 11 18	4.7	P	Roger, this is Friendship Seven. I have the Cape in sight down there. It looks real fine from up here.
03 11 24	0.9	CC	Rog. Rog.
03 11 26	1.0	P	As you know.
03 11 29	1.2	CC	Yea, verily, Sonny.
03 11 37	6.3	P	This is Friendship Seven, Flor-, I can see the whole state of Florida just laid out like on a map. Beautiful.
03 11 45	0.8	CC	Very good.

03 11 49	4.0	P	Even from this position out here, I can still see clear back to the Mississippi Delta.
03 12 32	7.8	P	This is Friendship Seven, checking down in Area Hotel on the weather and it looks good down that way. Looks like we'll have no problem on recovery.
03 12 41	1.6	CC	Very good. We'll see you in Grand Turk.
03 12 43	0.6	P	Yes, Sir.
03 12 48	10.5	P	In fact, I can see clear down, see all the islands clear down that whole chain from up here, can see way beyond them and area Hotel looks excellent for recovery.
03 13 13	1.4	CC	Friendship Seven, Bermuda Cap Com.
03 13 15	1.0	P	Go ahead, Bermuda.
03 13 17	4.3	CC	Cape recommends that you reenter on ASCS with manual for backup.
03 13 22	4.5	P	Roger, this is Friendship Seven. Understand recommend ASCS with manual backup.
03 13 29	30.3	P	This, this operation of ASCS has been very erratic. I have not been able to pin it down to any particular, one particular item. It went off one direction in yaw at one time; it went off the other the next time. I felt for a little while that pitch was drifting. It seemed to have a little stuck thrust in pitch at one time and I have to keep correcting that and that corrected itself and then I was off in roll just after I came off the dark side this time. Over.
03 14 00	3.3	CC	Roger, understand. I have all that down. Could you give us a blood pressure please?
03 14 03	1.2	P	Roger. Standby.
03 14 33	4.2	CC	Friendship Seven, can you give me both your primary and secondary oxygen?
03 14 38	5.4	P	Roger, primary oxygen is 62 [percent], secondary is 94 [percent].
03 14 45	0.4	CC	Roger
.03 15 18	5.7	P	This is Friendship Seven. I'm letting it drift a little bit to the right here to look up north.
03 15 30	1.4	CC	Friendship Seven, Bermuda Cap Com.
03 15 32	0.8	P	Go ahead Bermuda.
03 15 34	8.4	CC	Your fans inverter temperature is 205 [degrees] and your ASCS temperature is 195 [degrees].
03 15 44	2.3	P	Roger, sounds real good, Gus, fine.
03 15 47	0.9	CC	You're in good shape.
03 15 48	0.4	P	Roger.
03 16 18	3.3	CC	Friendship Seven, you want to go to HF now?
03 16 22	1.8	P	Roger, Friendship Seven. Going to HF.
03 16 37	2.5	CC	Friendship Seven, Bermuda Cap Com. How do you read on HF?
03 16 41	1.9	P	Bermuda, Friendship Seven. Loud and clear; how me?
03 16 44	3.2	CC	You're weak but readable, John.
03 16 48	0.5	P	Roger.
03 17 08	2.9	CC	Friendship Seven, Bermuda we have lost TM contact.
03 17 11	1.5	P	Roger, understand LOS.
03 17 43	3.5	P	This is Friendship Seven. I've drifted around almost to the 180° point again.
03 18 02	2.4	CC	Friendship Seven, Bermuda Cap Com. Are you still in contact?
03 18 05	6.9	P	Roger, this is Friendship Seven. I'm around at the 180 (degree) point. I'm facing the direction I'm traveling, Gus. Do you still receive? Over.
03 18 18	3.0	CC	Friendship Seven, Bermuda Cap Com. We do not read you,
03 18 22	1.0	P	Roger, Bermuda.

CANARY (THIRD ORBIT)

03 20 28	4.8	CT	Friendship Seven, Friendship Seven, this is CYI Com Tech, CYI Com Tech.

			Do you read? Over.
03 20 34	4.6	P	Hello, Canary Com Tech. On HF; how do you receive me? Loud and clear from you. Over.
03 20 41	3.5	CT	Friendship Seven, this is CYI Com Tech. Do you read? Over.
03 20 46	1.0	P	Roger, Canary.
03 20 49	4.7	CT	Rog, Roger. Read you 5 by, read you 5 by, here in the Canaries. Over.
03 20 54	3.9	P	Roger, Friendship Seven. I read you loud and clear on HF, also.
03 21 01	4.1	CC	Friendship Seven, Friendship Seven, this is Canary Cap Com. I read you loud and clear.
03 21 07	2.2	P	Roger, Canary. Loud and clear here also.
03 21 13	2.0	CC	Would you give me a brief station report? Over.
03 21 16	13.1	P	This is Friendship Seven. Fuel is 62-54. [percent]. Oxygen is 62-94, [percent], amps 24. Over.
03 21 30	5.9	CC	Understand that auto fuel is 63 [percent] and manual fuel is 94 [percent]. Over.
03 21 36	3.7	P	Negative. Manual fuel is 54 [percent], 54 [percent]. Over.
03 21 41	0.6	CC	54 [percent]. Over.
03 21 49	3.7	P	Canary, did you get manual fuel at 54 [percent] fiver four? Over.
03 21 54	2.0	CC	Roger. Manual fuel at 54 [percent].
03 22 02	2.7	CC	Friendship Seven, this is Canary Cap Com. How do you feel?
03 22 06	4.0	P	This is Friendship Seven. I feel fine; no effects whatsoever, none.
03 22 12	4.5	CC	Friendship Seven. Are you still seeing the particles around your capsule?
03 22 17	23.0	P	Negative. I don't seem to see them around here on this side. I saw a few, just a few just after I left Canaveral and turned around facing forward. They were coming toward me at that time. I was going, so I know that they were not coming from the capsule at all. I saw the particles in huge quantities at each sunrise so far. Over.
03 22 42	1.1	CC	Say again, Friendship Seven.
03 22 44	5.8	P	The particles I saw were mainly at sunrise each time around. Over.
03 22 51	3.6	CC	I understand. What control mode are you on?
03 22 55	2.9	P	This is Friendship Seven. I am in manual. Over.
03 22 59	1.1	CC	Roger. Understand.
03 23 14	4.8	CC	Friendship Seven, this is Canary Cap Com. The Aeromed wants to talk to you. Over.
03 23 17	1.4	P	Roger. Go ahead, Aeromed.
03 23 19	6.1	S	This is Canary surgeon. Are you having any nausea, or have you experienced any nausea at all during the entire flight?
03 23 25	5.0	P	This is Friendship Seven. Negative. I have felt perfectly normal during the whole flight. I feel fine. Over.
03 23 32	2.3	S	Very good. Back to Cap Com.
03 23 41	3.5	CC	Friendship Seven, Friendship Seven, do you want your inverter temperatures at this time?
03 23 45	1.9	P	Roger. This is Friendship Seven. Go ahead.
03 23 51	2.0	CC	Roger. We're turning you over to Canary Systems.
03 23 54	2.6	SY	Friendship Seven, this is Canary Systems. How do you read?
03 23 57	1.5	P	Go ahead, Canary Systems, go ahead.
03 24 00	17.9	SY	The systems on the ground look fairly normal at this time. The temperatures are a little bit on the high side but there's nothing critical showing up. The latest reports show that the inverters are about 200 [degrees] and 210 [degrees]; this does not seem to be critical. They should be holding at this level. Do you understand?
03 24 17	1.2	P	Roger. Friendship Seven.
03 24 21	5.9	SY	All other systems are as you reported to me.

ATLANTIC OCEAN SHIP (THIRD ORBIT)

03 24 43	7.2	CT	Friendship Seven, Friendship Seven, this is Atlantic Ship Cap, Com Tech. How do you read? Over.
03 24 51	3.6	P	Hello, Atlantic Cap Com. Read you loud and clear. How me? Over.
03 24 55	6.5	CT	Friendship Seven, Friendship Seven, this is Atlantic Ship Com Tech. We read you weak but broken, weak but broken.
03 25 01	10.3	P	Roger. Friendship Seven. Fuel 64-54 [percent], oxygen 62-94 [percent], amps 23. Over.
03 25 21	4.0	P	Hello, Atlantic Ship, Atlantic Cap Com. Friendship Seven, do you receive? Over.
03 25 34	9.0	CC	Friendship Seven, this is ATS Cap Com. How do you read me? Over. And what mode are you on for communications? Over.
03 25 43	8.3	P	This is Friendship Seven. I'm on HF at present time, on HF. I'll shift to UHF— is, if you're in solid contact. Over.
03 25 52	6.9	CC	Roger, Friendship Seven. We have TM contact, we have TM contact. Switch to UHF please.
03 26 00	1.4	P	Roger, switching to UHF.
03 26 12	1.8	P	Atlantic Ship, this is Friendship Seven. Over.
03 26 24	1.6	P	Hello, Atlantic Ship. Friendship Seven. Over.
03 26 29	5.5	CC	Friendship Seven, this is ATS Cap Com. Read only the last part of your transmission. Say again, please.
03 26 35	31.7	P	Atlantic Ship, This is Friendship Seven. Wish you would pass to Cape. I let the capsule drift around to the 180° position and I am having to reorient at present time. When I am all lined with the horizon and the periscope, my attitude indications now are way off. My roll indicates 30° right; my yaw indicates 35 [degrees] right; and pitch indicates plus 40 [degrees]; I repeat plus 40 [degrees] when I am in orbit attitude. Over.
03 27 11	0.8	P	Did you receive?
03 27 12	15.8	CC	Roger, Friendship Seven. I read you a little broken. You have discrepancies in attitudes of 30° right in roll, 35 [degrees] right in yaw and plus 40 [degrees] in pitch. Confirm, please. Over.
03 27 28	9.8	P	That is affirmative. I am realigning the capsule at present time and will cage and uncage the gyros before I go on the dark side. Over.
03 27 39	2.3	CC	Roger. We will standby.
03 28 58	2.5	P	This is Friendship Seven, ATS. Do you receive?
03 29 04	3.2	CC	Roger, Friendship Seven. Reading you loud and clear. Over.
03 29 07	4.5	P	This is Friendship Seven, in orbit attitude present time, caging gyros.
03 29 13	0.5	CC	Rog.
03 29 17	1.7	P	Gyros to cage, ready now.
03 30 09	5.7	P	This is Friendship Seven. Gyros are back on normal, going back to orbit attitude and will try ASCS. Over.
03 30 15	1.6	CC	Roger. Standing by.
03 30 57	5.0	P	Friendship Seven. Going to ASCS. Ready-now.
03 31 05	3.3	CC	Friendship Seven, this is ATS Cap Com. Do you read me? Over.
03 31 09	1.1	P	Roger, ATS.
03 31 15	3.9	CC	Friendship Seven, you are getting weak, you are getting weak. Over.
03 31 20	2.3	P	This is Friendship Seven. Did not read you, ATS.

ZANZIBAR (THIRD ORBIT)

03 31 23	10.9	CT	Friendship Seven, this is Zanzibar Com Tech, transmitting on HF UHF.
03 31 57	6.5	CT	Friendship Seven, Friendship Seven, this is Zanzibar Com Tech, transmitting on HF UHF. Do you read? Over.
03 32 04	6.9	P	Hello, Zanzibar Com Tech. Zanzibar Com Tech, Friendship Seven. Receive you weak but readable. How me? Over.

03 32 10	7.9	CT	Friendship Seven, Friendship Seven, this is Zanzibar Com Tech transmitting HF UHF. You're weak and garbled, weak and garbled. Do not copy. Over.
03 32 24	17.3	P	Roger. Friendship Seven. I receive you rather garbled, also. My condition is good. Fuel 5, correction, fuel 64-48 [percent]; oxygen 52-92 [percent]. Over.
03 32 44	9.9 (ATS)	CC	Friendship Seven, this is ATS Cap Com. I'm reading you very clear, very clear. Could you give me the reason for the errors in your attitudes, please?
03 32 54	41.6	P	This is Friendship Seven. That's a good question. I wish I knew, too. These errors have been off and on all during the flight. I have caged and recaged the gyros, caged and uncaged the gyros. I'm back in orbit attitude now but it is not, the attitude indicators are showing orbit attitude but it's not. By looking out at the horizon, I am about 20° right in roll and 20° too much on pitch down. I'm down to about probably 55° down in pitch by visual observation and about 20° right in roll. Yaw appears to be holding okay now. Over.
03 33 44	3.4 (ATS)	CC	Friendship Seven, ATS Cap Com.
03 34 14	2.5	P	Atlantic Ship, this is Friendship Seven. Over.
03 35 58	3.0	P	This is Friendship Seven, approaching sunset.
03 36 09	6.7	CT	Friendship Seven, Friendship Seven, this is Zanzibar Com Tech transmitting on HF UHF. Please acknowledge on HF. Over.
03 36 49	18.6	P	This is Friendship Seven recording. A lot of dirt on the windows from the retrofire and lot of stuff on here looks like, ah, we might have smashed some bugs even on the way up off the pad. Looks like blood on the outside of the window, maybe. It makes it real, very difficult to observe anything when they get around on the sun side.
03 37 32	8.3	P	Friendship Seven. The sun is going down again now. Coming off automatic in yaw, and yawing a little bit to the left to observe it.
03 38 05	1.7	P	Mark, the sun is down.
03 38 29	18.3	P	And can see little or nothing of zodiacal light at the moment.
03 39 21	10.3	P	This is Friendship Seven, flying with yaw handle pulled. Controlling on manual yaw.
03 39 41	18.1	P	The way the horizon looks is a very orange band. Just as the sun goes down and extends way off either side, probably 45° each side of the sun, comes up into a lighter yellow, then a very deep blue, then a very light blue, on up to the black of the sky.
03 40 06	3.2	P	Once again I can see lightning flashing under me, very clearly.
03 40 36	7.4	CT	Friendship Seven, Friendship Seven, this is Zanzibar Com Tech, transmitting on HF UHF. Please acknowledge on HF. Over.
03 40 44	4.7	P	Hello, Zanzibar Com Tech. Friendship Seven on HF. Do you receive me now. Over.
03 40 50	4.2	CC	Friendship Seven, this is Zanzibar Cap Com reading you weak but garbled.
03 40 58	3.0	P	Roger. Understand reading me weak and garbled.
03 41 02	16.8	P	This is Friendship Seven. I took the capsule off of automatic in yaw only to go left to look at the sunset. It's back on automatic at present time in all three axes; I'm backing it up with. manual. Over.
03 41 20	7.3	CC	Roger, Friendship Seven, understand you're in automatic control in all three axes, backing up with manual. Is this affirmative. Over.
03 41 28	1.3	P	That is affirmative,
03 41 32	18.5	P	There's quite a big storm area under me. It must extend for, I see lightning flashes, as far, way off on the horizon to the right. I also have them almost directly under me here. They show up very brilliantly here on the dark side at night. They're just like firecrackers going off. Over.
03 41 59	4.6	P	Zanzibar, this is Friendship Seven. Did you get my fuels when I reported them? Over.
03 42 05	1. 6	CC	Repeat your last transmission. Over.

03 42 07	13.2	P	Roger. This is Friendship Seven. Fuel is 60-55 [percent], oxygen is 60-92 [percent], Amps are 22. Over.
03 42 28	5.5	p	This is Friendship Seven. All voltages are 25, or above. Over.
03 42 53	12.3	P	This is Friendship Seven at 3 plus 42. We should be just about over Johannesburg, I cannot see anything of southern Africa on this pass. Over.
03 43 06	2.2	CC	Zanzibar Cap Com. Do you read? Over.
03 43 12	2.8	P	Roger, Zanzibar. I'm reading you weak but readable.
03 43 46	16.4	P	Zanzibar Cap Com, Friendship Seven on HF. Can see long streaky clouds down below as we, going off to my right up into sort of a general weather pattern. That's 3 plus 44.
03 44 45	32.1	P	This is Friendship Seven. An observation on control system operation: It appears that whenever I go off on manual, or fly-by-wire and maneuver for any lengthy period of time, that somehow we induce considerably, considerable errors into the gyro system. I come up with all kinds of attitudes. That time for instance, when I yawed around 180°, and held for a little while, and came back again, I had errors of 30° in roll, 35 [degrees] in yaw and plus 40 [degrees] in pitch.
03 45 23	5.9	P	This is Friendship Seven, going on manual pitch.
03 46 08	22.5	P	This is Friendship Seven. Several times I have felt that I had a partially stuck thruster or one that was just partly operating in pitch down and just then again I had a pitch up rate going and all at once I felt a down thrust and it pitched down on me. Again there, it repeated again that time, so I think I have a stuck pitch thruster occasionally.

INDIAN OCEAN SHIP (THIRD ORBIT)

03 47 03	4.6	CT	Friendship Seven, Friendship Seven, this is Indian Com Tech. How do you read? Over.
03 47 09	2.1	P	Hello, Indian Com Tech, Friendship Seven. Over.
03 47 14	8.2	CT	Friendship Seven, this is Indian Com Tech. I read you very weak, very weak, very garbled. Am turning over to Cap Com on UHF. Over.
03 47 23	1.3	P	Roger. Friendship Seven.
03 47 30	4.5	CC	Friendship Seven, Friendship Seven, this is Indian Cap Com on UHF. Do you read? Over.
03 47 34	2.8	P	Roger, Indian Cap Com. I read you loud and clear. How me?
03 47 39	3.6	CC	About 3 by 3. You're coming in stronger. Over.
03 47 42	10.1	P	Roger. Friendship Seven. Fuel is 60-45 [percent], oxygen is 60-92 [percent], amps 23. Over.
03 47 54	2.3	CC	Say again your oxygen, Over.
03 47 58	5.7	P	Roger. Oxygen is 60-92 [percent]. Over.
03 48 04	1.5	CC	Roger, understand.
03 48 14	4.1	CC	Friendship Seven, this is Indian Cap Com. What is your control mode? Over.
03 48 19	12.9	P	This is Friendship Seven. I'm on ASCS but it is operating very erratically. I'm backing it up with manual at the present time. I'm trying to get it set up so it will be in a decent ASCS attitude for retrofire. Over.
03 48 33	1.4	CC	Roger, understand.
03 48 36	5.8	P	This is Friendship Seven. I can control it manually. I'll back it up manually and take over if I need to. Over.
03 48 42	1.3	CC	Roger, understand.
03 48 51	5.2	CC	Friendship Seven, this is Indian Cap Com. I have your retrosequence times. Are you prepared to copy? Over.
03 48 57	2.2	P	This is Friendship Seven. Standby one.
03 49 00	0.7	CC	Standing by.
03 49 25	1.2	P	Friendship Seven. Go ahead.
03 49 27	8.2	CC	Roger. Area 3 Delta, Area 3 Delta is 4 hours, 12 minutes, 32 seconds.

03 49 36	0.8	P	Roger.
03 49 40	4.2	P	Area 3 Delta is 04 plus 12 plus 32. Is that affirm?
03 49 45	10.2	CC	Affirmative. Area 3 Echo, Area 3 Echo is 4 hours, 22 minutes, 12 seconds. I say again; 4 hours, 22 minutes, 12 seconds. Over.
03 49 55	4.6	P	Roger. Area 3 Echo is 04 plus 22 plus 12.
03 50 01	9.9	CC	Roger. Area Hotel is 4 hours, 32 minutes, 37 seconds. Say it again; 4 hours, 32 minutes 37 seconds. Over.
03 50 11	3.5	P	Roger. 04 plus 32 plus 38. I have correction, plus 37; I have 38 set on my retro sequence because of error in my clock of 1 second. Over.
03 50 15	2.7	CC	That is affirmative. We assume you have an error in your clock. Over.
03 50 20	9.7	P	One second error, that is affirm. Request you confirm with Cape that I have 04 plus 32 plus 38 as correct retrograde time. Over.
03 50 40	1.2	CC	Roger. That's affirmative.
03 50 49	8.3	P	This is Friendship Seven. The ASCS is drifting again. I'm indicating 25° right in yaw. Over.
03 50 59	1.0	CC	Roger, understand.
03 51 05	6.6	P	This is Friendship Seven. I have almost continuous cloud cover under me as far as I can see in every direction. Over.
03 51 13	1.2	CC	Roger, understand.
03 51 18	3.5	CC	Friendship Seven, this is Indian Cap Com. Surgeon would like to talk to you. Over.
03 51 22	1.6	P	Roger. This is Friendship Seven.
03 51 25	4.6	8	This is the surgeon here. Have you switched to secondary oxygen for any reason during this hop? Over.
03 51 29	7.6	P	Negative. I don't quite understand the decrease in secondary either unless it's the bottles are beginning to cool down, but they shouldn't cool that much.
03 51 38	3.1	S	We're reading 90 [percent] on the TM. How about you?
03 51 41	1.5	P	I am reading 90 [percent] also.
03 51 44	2.5	S	Somewhere about 65 [percent]. This is Indian Surgeon. Out.
03 51 47	14.1	P	Roger. This is Friendship Seven. This thing is slipping in and out of orientation mode and wasting fuel at present time. I'm just going to try and hold it on orbit attitude manually. Over. Or on fly-by-wire. Over.
03 52 02	2.0	CC	Roger. Manually by fly-by-wire. Over.
03 52 04	0.5	P	Roger.
03 52 18	7.0	P	This is Friendship Seven. Checking different control modes on fly-by-wire. I have no low thrust to the right. Over.
03 52 27	2.3	CC	Roger, understand no low thrust to the right.
03 52 29	0.4	P	Roger.
03 52 32	3.4	P	I do have low thrust in all other axes. Over.
03 52 37	1.2	CC	Roger, understand.
03 53 04	3.0	P	This is Friendship Seven. Pitching up for star observation.
03 53 10	1.9	CC	Roger. Report anything you see. Over.
03 53 12	0.8	P	Roger.
03 54 31	3.8	CC	Friendship Seven, this is Indian Cap Com. Are you able to see anything? Over.
03 54 36	11.7	P	This is Friendship Seven. Affirmative. I have Orion right in the middle of the window at present time and makes a good one to hold attitude on. I have, I am using it as horizon reference at the moment.

MUCHEA (THIRD ORBIT)

03 55 05	3.4	CT	Friendship Seven, this is Muchea Com Tech. How do you read? Over.
03 55 09	4.5	P	Hello, Muchea Com Tech. Loud and clear. Standby one. I'm right in the

			middle of operation here.
03 55 29	3.9	CT	Friendship Seven, Friendship Seven, Muchea Com Tech. Do you read? Over.
03 55 45	3.2	P	Hello, Muchea Com, Tech, Roger. Read you loud and clear. How me?
03 55 59	3.8	CT	Friendship Seven, Friendship Seven, Muchea Com Tech. Do you read? Over.
03 56 17	5.3	P	Hello, Muchea Com Tech, Muchea Com Tech. Roger. Friendship Seven. Loud and clear. How me?
03 56 23	0.8	CT	Over.
03 56 25	1.5	P	Roger. This is Friendship Seven.
03 56 30	5.6	CT	Roger, Friendship Seven. I'm reading you 3 by 3 on HF. Over. Would you call again?
03 56 37	3.7	P	Roger, Friendship Seven, reading you loud and clear on HF. Over.
03 56 52	4.3	CT	Friendship Seven, Friendship Seven, Muchea, Com Tech, say again.
03 56 58	3.3	P	Roger, Muchea Com Tech. Friendship Seven. Loud and clear. How me?
03 57 11	3.5	CT	Friendship Seven, Friendship Seven, this is Muchea Com Tech. Do you read?
03 57 15	3.9	P	Roger, Muchea Com Tech, Friendship Seven. Read you loud and clear. How me?
03 57 20	3.7	CC	Roger, Friendship Seven. You are 3 by 3. Go ahead to Cap Com.
03 57 24	4.0	P	Roger. Hello Cap Com, Friendship Seven. How are things going?
03 57 36	3.3	P	Hello, Muchea Cap Com, Muchea Cap Com, Friendship Seven. Over.
03 57 48	2.6	P	Hello, Muchea Cap Com, Muchea Cap Com. Over.
03 57 53	1.2	CC	How me? Over.
03 58 01	2.7	CC	Friendship Seven, Muchea Cap Com. How now? Over.
03 58 04	3.6	P	Muchea Cap Com, Friendship Seven. Loud and clear. How me?
03 5909	3.8	CC	Roger. You are coming through rather weak. Do you want to switch to UHF? Over.
03 58 13	1.7	P	Roger. Switching to UHF
03 58 24	2.8	CC	Friendship Seven, Muchea Cap Com, how do you read? Over.
03 58 27	2.6	P	Roger, Muchea. Loud and clear, how me? Over.
03 58 33	3.8	P	Hello, Muchea, Friendship Seven. Loud and clear, how me? Over.
03 58 38	3.0	CC	Muchea Cap Com. Give us 30-minute report.
03 58 43	15.1	p	Roger. This is Friendship Seven. 30-minute report: ASCS is major item, still not operating properly. I am on fly-by-wire at present time. I have no low thrust to the right fly-by-wire. Over.
03 58 58	3.5	CC	Roger, understand no low right thruster on fly-by-wire.
03 59 02	2.2	P	That is affirmative. Got it?
03 59 06	4.8	CC	We would like to send you a Z and R Cal sometime during your 30-minute report ...
03 59 11	0.9	P	Roger. Fine.
03 59 13	2.8	CC	You want to start down the 30-minute report there?
03 59 16	3.6	P	This is Friendship Seven. In 45 more seconds I would like to have you send a message for me, please. Over.
03 59 23	13.9	P	I want you to send a message to the Director, to the Commandant, U.S. Marine Corps, Washington. Tell him I have my 4 hours required flight time in for the month and request flight chit be established for me. Over.
03 59 39	0.9	CC	Roger. Will do.
03 59 47	1.1	CC	Think they'll pay it?
03 59 49	3.0	P	I don't know. Gonna find out.
03 59 51	3.7	CC	Roger. Is this flying time or rocket time?
03 59 53	1.4	P	Lighter than air, buddy.
03 59 59	0.3	CC	Rog.
04 00 03	2.3	CC	We're sending you a Z and R Cal, hear.
04 00 07	0.5	P	Very well.
04 00 09	0.8	CC	Coming now.
04 00 11	3.4	CC	Are you going to start down on this procedure, or this 30-minute stuff?
04 00 15	36.3	P	Yes, all the fuse switches are still in the same position, Gordo. I haven't

changed any of those. Squib is armed, Auto Retrojettison is off. I am on fly-by-wire, auto, and gyro normal. All "T" handles are in. The sequence panel is normal except for Landing Bag in the off position. The attitude indicator has been rather erratic, they drift very rapidly sometimes. Especially when I maneuver myself and then come back onto ASCS. It seems they have been thrown way off at that time.

04 00 51	0.4	CC	Roger.
04 00 53	10.6	P	My retrograde time 04 plus 32 plus 38 for capsule time which corrects for 1 second error in this clock,
04 01 09	9.3	CC	On that, MCC recommends you change that to 04 32 37 shortly after you get done with these others.
04 01 14	55.6	P	Roger. Okay, cabin pressure is still holding at 5.5. It's been there ever since we left. 90 [percent] on cabin temp, relative humidity is back up again now to about 36 [percent], coolant quantity is down a little bit too, we may have had a small water leak some place since we have that much increase in humidity, cause it was down around 20 [percent]. Our coolant quantity, though, is down around 62 [percent] now. The suit temperature is 70 [degrees] inlet; cabin suit pressure is 5.8. Steam temperature is 4 point, is 48 [degrees]. Oxygen is 60-90[percent]. Amps are 24, ASCS is 11, about 115 now. Fantastic. Over.
04 02 10	3.2	CC	Roger. All your switches outboard?
04 02 13	7.9	P	That's affirmative. On the right side, everything is outboard except the fuel quantity warning light which is on. I have that switch inboard to cut the audio.
04 02 22	6.2	P	The only 2 switch fuses in the center off position, on the right, are Retro Jettison and Retro Manual. Over.
04 02 29	1.7	CC	Roger. Comfort control settings?
04 02 32	2.5	P	Control settings on the water?
04 02 36	0.3	CC	Rog.
04 02 44	17.4	P	Water on, on cabin temperature is setting number 2. Setting on the suit temp is beyond the 1.7 mark. I repeat, beyond the 1.7 mark which is the maximum setting.
04 02 55	4.9	CC	Roger. I understand. What is your opinion on the gyro problem, John?
04 03 00	11.3	P	Well, I don't know. I want to start lining up just as carefully as I can here in a minute and see whether the scanners will pick it up and correct it in so that we have a good retrofire attitude. If it is not, I'll align it myself. Over.
04 03 12	4.8	CC	Roger. Do you have your three Dog, Easy, and Hotel times from Indian Ocean Ship?
04 03 17	1.7	P	Yes I did. I got these okay.
04 03 19	15.3	CC	And, now I gave you the recommended change in the retro clock to 04 32 37. On your retro, using ASCS, you'll be using high torque thrusters for retrofire mode there. What do you say about retroing by ASCS and backing up by fly-by-wire?
04 03 36	6.5	P	Well-well, you can't, you couldn't do it on ASCS and fly-by-wire. You mean on manual.
04 03 43	4.9	CC	No, I meant you could go ASCS by fly-by-wire and back it up on manual proportional.
04 03 48	17.6	P	Yes. If the ASCS appears to be programming and holding a good orbit attitude I'll let it go on ASCS and back it up with manual. If not, if it appears that the gyros are cooked as they were a little while ago, then I'll just stay on manual retrofire, I think and let it go at that.
04 04 07	2.9	CC	Are you in manual proportional or fly-by-wire now?
04 04 10	1.5	P	I'm in fly-by-wire now.
04 04 21	6.4	P	This is Friendship Seven. I am going to as near orbit attitude as I can establish here on the dark side.

WOOMERA (THIRD ORBIT)

04 04 40	4.0	CC	Friendship Seven, Woomera Cap Com. We have contact UHF and TM solid.
04 04 44	2.3	P	Roger, Woomera, loud and clear.
04 04 49	8.9	P	Fuel is 52-45 [percent], oxygen is 60-90 [percent], amps 25. Over.
04 05 06	1.5	P	Hello, Woomera, did you receive? Over.
04 05 08	6.4	CC	Roger, Friendship Seven. Surgeon recommends that if you haven't eaten that you eat in the near future.
04 05 18	12.3	P	This is Friendship Seven. Negative. I did not eat on that last round because of, I was busy with the ASCS problem. Over.
04 05 30	4.1	CC	Friendship Seven, let's have your fuel readings. Over.
04 05 35	1.4	P	Say again, Woomera.
04 05 38	2.1	CC	Let's have your fuel readings. Over.
04 05 41	6.4	P	Roger. Fuel readings are 50-45 [percent], correction 52-45 [percent], over.
04 05 49	3.0	CC	Roger, . . . temperature. Over.
04 05 55	1.5	P	Say again please. Over.
04 05 58	3.9	CC	Friendship Seven, you are fading . . . temperature. Over.
04 06 02	4.0	P	Roger, steam temperature is 49-49 [degrees]. Over.
04 06 08	1.6	CC	Roger, Friendship Seven.
04 06 10	4.3	P	This is Friendship Seven. Turning more water on on that one, see if I can cool it down a little more.
04 07 30	7.6	P	This is Friendship Seven. I can notice a little crackling on fly-by-wire switches when I operate fly-by-wire; it comes in on the head set.
04 13 41	16.8	P	This is Friendship Seven. Turned around, yawed 180 [degrees] to see the sunrise here, and also to see these little, these little gadgets here that I don't know what they are.
04 14 04	7.9	P	They do not seem to be coming from the capsule at all. There are too many of them. They're all spread out all over the place; it looks like they're some of them might be miles away.
04 15 23	5.4	CT	This is Com Tech, Canton Com Tech. Do you read? Over.
04 15 28	3.2	P	Go ahead, Canton.
04 15 33	5.9	CT	Com Tech, Roger, read you loud and clear. Friendship Seven, this is Canton Com Tech, you are weak but readable. Standby for Cap Com.
04 15 40	0.6	P	Roger.
04 15 42	2.5	CC	This is Canton Cap Com. Over.
04 15 45	5.9	P	Canton Cap Com, standby I'll give you a report in a minute here. I'm maneuvering back into retro attitude.
04 15 52	1.2	CC	OK. Standing by.
04 18 02	6.5	CC	Friendship Seven, this is Canton Cap Com. We are not receiving your transmissions and we have not lost your contact. Over.
04 18 09	24.2	P	Roger, Canton. I was busy maneuvering here. I did not give you your report yet, was getting lights set up and trying to stow everything for retro fire. I have 45-45 [percent] on fuel, oxygen is 60-90 [percent], amps is 23. My retrograde time is still set at 04 plus 32 plus 38. Over.
04 18 35	3.2	CC	Roger, Friendship Seven. Canton standing by.
04 18 38	0.4	P	Roger.
04 19 19	6.5	P	This is Friendship Seven, Canton. I am getting in orbit attitude so I can cage and uncage the gyros again. They're off again. Over.
04 19 27	1.3	CC	Roger, Friendship Seven.
04 19 29	1.5	P	Request you notify Cape.
04 19 34	4.5	CC	Friendship Seven, this Cap Com. I did not read your last transmission. Would you repeat please?
04 19 39	16.3	P	Roger, Canton. Please notify Cape that I am indicating a roll 10° right, yaw 10° right, and pitch a plus 15 [degrees] when I'm in orbit attitude on the

window and the scope. Over.

HAWAII (THIRD ORBIT)

04 21 00	2.0	CT	Friendship Seven, Hawaii Com Tech. How do you read? Over.
04 21 04	2.1	P	Loud and clear, Hawaii Com Tech. How me?
04 21 09	2.4	CT	Roger. Reading you loud and clear on HF.
04 21 13	6.2	p	Roger, this is Friendship Seven. Caging gyros and uncaging.
04 21 43	8.5	P	Hawaii Com Tech, Cap Com, this is Friendship Seven. Fuel 43-45 [percent], oxygen is 60-90 [percent], amps is 23. Over.
04 21 56	2.0	P	Hello, Hawaii Com Tech. Do you receive? Over.
04 21 59	5.3	CT	Friendship Seven, Hawaii Com Tech. Roger. Transfer from HF, would you go to UHF? Over.
04 22 05	1.3	P	Roger. Going UHF.
04 22 30	2.7	CT	Friendship Seven, do you read on UHF? Over.
04 22 35	2.0	P	Roger, read UHF loud and clear.
04 22 38	3.9	CC	Friendship Seven, Hawaii Cap Com. Reading you loud and clear on UHF.
04 22 45	18.0	CC	Friendship Seven, we have been reading an indication on the ground of segment 51, which is Landing Bag Deploy. We suspect this is a erroneous signal. However, Cape would like you to check this by putting the Landing Bag switch in auto position, and see if you get a light. Do you concur with this? Over.
04 23 09	6.6	P	Okay. If that's what they recommend, we'll go ahead and try it. Are you ready for it now?
04 23 16	0.9	CC	Yes, when you're ready.
04 23 17	6.8	P	Roger. Negative, in automatic position did not get a light and I'm back in off position now. Over.
04 23 25	4.9	CC	Roger, that's fine. In this case, we'll go ahead, and the reentry sequence will be normal.
04 23 31	1.8	P	Roger, reentry sequence will be normal.
04 23 34	3.1	CC	Friendship Seven, have you completed your pre-retro check list?
04 23 38	4.3	P	This is Friendship Seven. Going to pre-re, check list at present time.
04 23 43	1.1	CC	Roger. I'll standby.
04 25 05	1.4	CC	Friendship Seven, Hawaii Cap Com.
04 25 08	0.7	P	Go ahead, Hawaii.
04 25 10	2.9	CC	Have you completed pre-retro check list at this time?
04 25 13	2.2	P	That's affirmative. I'm just now completing it.
04 25 17	10.6	CC	Roger. Can you comment again on this, the attitude system with respect to the, respect to the visual reference, how you feel about this at this time?
04 25 28	12.5	P	Well, it's into, it's in and out of orientation mode right now and is wasting fuel, in yaw. I may have to cut it in yaw, I don't know. It appears to be correcting, though, at the present time or maybe the scanners are correcting it okay now.
04 25 42	1.6	CC	Roger, what mode are you in now?
04 25 44	1.9	P	I'm in ASCS, automatic.
04 25 47	0.4	CC	Understand.
04 25 51	7.4	CC	Did you understand the Cape would like you to change your clock by one second, to 04 32 37?
04 25 58	1.7	P	Negative. I'll change it right now.
04 26 03	4.3	P	Okay. Time is now 04 32 and 37.
04 26 08	1.2	CC	We confirm with TM readout.
04 26 10	0.5	P	Roger.
04 26 13	8.7	CC	Everything looks good on the ground. The inverter temperatures are a little high; 225° on the 150, 212° on the 250. Everything else looks pretty good.
04 26 22	0.5	P	Roger.

04 26 28	2.7	CC	Surgeon would like to know if you're still comfortable.
04 26 31	5.3	P	Roger, I'm in very good shape. I'll go through exercise bit in just a minute here, as soon as I get done with check list.
04 26 05	7.8	P	Okay, going through light test. Okay, checks okay.
04 27 38	2.4	P	Okay, 5 minutes to retrograde, light is on.
04 27 43	1.7	CC	Rog, 5 minutes to retrograde, light on.
04 27 54	18	CC	TM is breaking up now. Friendship Seven, would you like a G.m.t. time hack?
04 27 58	0.8	P	Roger, please.
04 28 00	4.8	CC	On my mark G.m.t. will be 19 plus 15 plus 45—MARK.
04 28 08	0.6	P	Roger.
04 29 10	7.1	CC	GET on my mark will be 04 plus 2
04 28 18	2.5	P	Say again capsule elapsed time please. Over.
04 28 21	6.9	CC	Give you a new hack. On my mark it will be 04 plus 28 plus 35, six seconds.
04 28 29	15.5	P	Roger. Let me give you a hack and figure a new retrograde time from the Cape. My time from launch will read 04 plus 28 plus 45 on my mark. Standby, 2, 3, 4, MARK.
04 28 49	4.5	CC	Friendship Seven, Hawaii Cap Com. You were breaking up at the last, I could not read your time hack.
04 28 56	13.4	P	Roger, this is Friendship Seven. I'll give you another time hack at 04 plus 26 plus 10. Standby, 7, 8, 9, MARK.
04 29 14	1.4	P	Hawaii, did you receive? Over.
04 29 29	1.3	P	Hello Hawaii, did you receive?

CALIFORNIA (THIRD ORBIT)

04 29 56	2.1	P	Hello California, Friendship Seven. Over.
04 30 11	2.3	p	Hello California Cap Com, Friendship Seven. Over.
04 30 20	2.2	P	Hello California Cap Com. Friendship Seven. Over.
04 30 32	2.8	CC	Friendship Seven, this is California Cap Com. Read you loud and clear, how me?
04 30 35	3.7	P	California Cap Com, Friendship Seven, UHF. Do you receive now? Over.
04 30 41	2.8	P	Hello California Cap Com, Friendship Seven, UHF. Over.
04 30 45	2.4	CC	Friendship Seven, California Cap Com. How do you read? Over.
04 30 49	2.5	P	This is Friendship Seven. Loud and clear. How me? Over.
04 30 56	2.4	P	Hello California Cap Com, Friendship Seven. Over.
04 31 06	1.8	P	California Cap Com, Friendship Seven. Over.
04 31 09	3.7	CT	Seven, this is California Com Tech, California Com Tech. How do you read? Over.
04 31 13	2.9	P	This is Friendship Seven. Loud and clear. How me? Over.
04 31 18	4.1	CT	Roger, Friendship Seven, this is California Com Tech. Reading you loud and clear on UHF.
04 31 21	16.7	P	Roger, this is Friendship Seven. Let me give you my time, my capsule elapsed time is 04 plus 31 plus 35 on my mark. 2, 3, 4, MARK. Will you relay that immediately to Cape? I think we're several seconds off. Over.
04 31 40	5.5	CC	Roger, we have you on that. Will give you the count down for retro-sequence time, John. You're looking good.
04 31 46	3.7	P	Roger. We only have 50 seconds to retrograde. Over.
04 31 50	2.7	CC	John, I'll give a mark. 45, MARK.
04 31 53	0.3	P	Roger.
04 31 57	2.8	P	I'm on ASCS and backing it up manual. Over
.04 32 02	0.5	CC	Roger, John.
04 32 07	1.6	P	My fuel is 39 [percent]
04 32 09	1.0	CC	Thirty seconds, John.
04 32 11	1.6	P	Roger. Retro-warning is on.
04 32 13	0.3	CC	Good.

04 32 15	4.1	CC	John, leave your retropack on through your pass over Texas. Do you read?
04 32 19	0.6	P	Roger.
04 32 23	1.5	CC	15 seconds to sequence.
04 32 25	0.5	P	Roger.
04 32 28	0.2	CC	10.
04 32 32	5.2	CC	5, 4, 3, 2, 1, MARK.
04 32 39	2.0	P	Roger, retro sequence is green.
04 32 42	2.0	CC	You have a green. You look good on attitude.
04 32 44	1.5	P	Retro attitude is green.
04 32 50	1.2	CC	Just past 20.
04 32 52	0.4	P	Say again.
04 32 53	0.2	CC	Seconds.
04 32 55	0.4	P	Roger.
04 33 02	5.2	CC	5, 4, 3, 2, 1, fire.
04 33 09	2.4	P	Roger, retros are firing.
04 33 12	0.9	CC	Sure, they be.
04 33 15	3.1	P	Are they ever. It feels like I'm going back toward Hawaii.
04 33 19	1.9	CC	Don't do that, you want to go to the East Coast.
04 33 23	2.2	P	Roger. Fire retro light is green.
04 33 26	0.7	CC	All three here.
04 33 28	0.5	P	Roger.
04 33 32	4.1	P	Roger, retros have stopped. A hundred ....
04 33 37	2.5	CC	Keep your retro pack on until you pass Texas.
04 33 40	1.0	P	That's affirmative.
04. 33 41	0.3	CC	Check.
04 33 47	2.9	CC	Pretty good looking flight from what all we've seen.
04 33 51	3.4	P	Roger, everything went pretty good except for all this ASCS problem.
04 33 55	2.8	CC	It looked like your attitude held pretty well. Did you have to back it up at all?
04 33 57	3.9	P	Oh, yes, quite a bit. Yeah I had a lot of trouble with it.
04 34 04	2.1	CC	Good enough for Government work from down here.
04 34 06	2.2	P	Yes, sir, it looks good, Wally. We'll see you back East.
04 34 08	0.2	CC	Rog.
04 34 09	0.6	P	All right, boy.
04 34 11	1.3	P	Fire Retro is green.
04 34 14	0.5	CC	Roger.
04 34 15	2.2	P	Jettison retro is red. I'm holding onto it.
04 34 18	0.5	CC	Good head.
04 34 28	2.4	P	I'll tell you, there is no doubt about it when the retros fire.
04 34 32	1.4	CC	Gathered that from your comments.
04 34 39	35.1	P	Everything is looking good, I'll give you a fast readout here. Fuel is 29-27 [percent]. The cabin pressure holding 5.5, cabin air is 88 [degrees], relative humidity is 33 [percent]; coolant quantity , 58 [percent], temperature is 71 [degrees], suit temperature is 71 [degrees], suit pressure is 5.8, steam temperature is 53 [degrees] in the suit, oxygen is, primary 60 [percent], 89 [percent] on secondary.
04 35 17	4.3	CC	Looks pretty good on this end. How did the attitude seem to hold? Did you have any diversions in yaw at all?
04 35 21	5.3	P	Negative, very close. I backed it up and worked right along with the ASCS and it looked like it held right on the money.
04 35 27	2.9	CC	Roger, we didn't notice any particular disparity here.
04 35 29	4.0	P	Roger, good. Do you have a time for going to Jettison Retro? Over.
04 35 33	1.4	CC	Texas will give you that message. Over.
04 35 35	0.3	P	Roger.
04 35 39	6.7	P	This is Friendship Seven, cutting yaw on automatic and I'll control that manually; it keeps banging in and out of orientation.

04 35 46	1.3	CC	Roger, Friendship Seven.
04 35 49	1.7	P	Hello, Texas. Friendship Seven. Over.
04 35 56	1.9	CC	Friendship Seven, Cal Cap Com. Do you read?
04 35 58	0.6	P	Roger.
04 35 59	3.3	CC	Consideration about leaving retropack on, they will inform you over Texas.
04 36 03	0.5	P	Roger.
04 36 05	1.2	P	Roger, over the Coast.
04 36 07	1.2	CC	Roger, clear blue here.
04 36 09	0.4	P	Yes, sir
04 36 28	5.6	P	This is Friendship Seven. Can see El Centro and the Imperial Valley down there; Salton Sea very clear.
04 36 34	2.4	CC	It should be pretty green; we've had a lot of rain down here
04 36 37	0.4	P	Yes, sir.
04 36 47	2.5	CC	Do you notice any contrast over the coastline, John?
04 36 50	0.8	P	Say again.
04 36 51	1.8	CC	How about contrast over the coastline?
04 36 53	0.7	P	Negative
04 36 55	0.4	CC	Roger.
04 36 59	9.6	P	There is quite a bit of cloud cover down in this area. I can, right on track, I can only see certain areas. I can see quite a bit on up to the north, however.
04 37 16	2.4	P	This is Friendship Seven, going to manual control.
04 37 19	1.3	CC	Roger, Friendship Seven.
04 37 21	2.7	P	This is banging in and out here; I'll just control it manually.
04 37 23	0.4	CC	Roger.
04 37 46	3.1	CC	Friendship Seven, Guaymas Cap Com, reading you loud and clear.
04 37 49	2.1	P	Roger, Guaymas, read you loud and clear also.

TEXAS (THIRD ORBIT)

04 38 04	4.0	CT	Friendship Seven, Friendship Seven, this is Texas Com Tech. Do you read? Over.
04 38 08	1.3	P	Roger, Texas, go ahead.
04 38 11	3.9	CT	Roger. Reading you 5 square. Standby for Texas Cap Com.
04 39 14	0.4	P	Roger.
04 38 23	23.8	CC	This is Texas Cap Com, Friendship Seven. We are recommending that you leave the retropackage on through the entire reentry. This means that you will have to override the 0.05g switch which is expected to occur at 04 43 53. This also means that you will have to manually retract the scope. Do you read?
04 38 47	4.0	P	This is Friendship Seven. What is the reason for this? Do you have any reason? Over.
04 38 51	3.6	CC	Not at this time; this is the judgment of Cape Flight.
04 38 56	2.6	P	Roger. Say again your instructions please. Over.
04 38 59	22.1	CC	We are recommending that the retropackage not, I say again, not be jettisoned. This means that you will have to override the 0.05g switch which is expected to occur at 04 43 53. This is approximately 4½ minutes from now. This also means that you will have to retract the scope manually. Do you understand?
04 39 23	9.7	P	Roger, understand. I will have to make a manual 0.05g entry when it occurs, and bring the scope in manually. Is that affirm?
04 39 33	2.5	CC	That is affirmative, Friendship Seven.
04 39 37	0.6	P	Roger.
04 39 40	3.6	P	This is Friendship Seven, going to reentry attitude, then, in that case.
04 39 58	3.8	CC	Friendship Seven, Cape flight will give you the reasons for this action when you are in view.

04 40 04	2.6	P	Roger. Roger. Friendship Seven.
04 40 07	2.5	CC	Everything down here on the ground looks okay.
04 40 10	1.5	P	Roger. This is Friendship Seven.
04 40 12	1.4	CC	Confirm your attitudes.
04 40 14	0.4	P	Roger.

CANAVERAL (THIRD ORBIT)

04 40 21	1.7	CC	Friendship Seven, this is Cape. Over.
04 40 23	1.5	P	Go ahead, Cape. Friendship Seven,
04 40 25	4.9	CC	Recommend you go to reentry attitude and retract the scope manually at this time.
04 40 30	1.9	P	Roger, retracting scope manually.
04 40 34	14.6	CC	While you're doing that, we are not sure whether or not your landing bag has deployed. We feel it is possible to reenter with the retropackage on. We see no difficulty at this time in that type of reentry. Over.
04 40 49	1.6	P	Roger, understand.
04 41 08	1.4	CC	Seven, this is Cape. Over.
04 41 10	1.5	P	Go ahead, Cape. Friendship Seven.

CANAVERAL (REENTRY)

04 41 13	5.4	CC	Estimating 0.05g at 04 44.
04 41 19	0.6	P	Roger.
04 41 21	3.0	CC	You override 0.05g at that time.
04 41 29	2.7	P	Roger. Friendship Seven.
04 41 31	13.2	P	This is Friendship Seven. I'm on straight manual control at present time. This was, still kicking in and out of orientation mode, mainly in yaw following retrofire, so I am on straight manual now. I'll back it up.
04 41 43	0.8	CC	... on reentry.
04 41 45	0.9	P	Say again.
04 41 48	0.6	CC	Standby.
04 41 51	6.2	P	This is Friendship Seven. Going to fly-by-wire. I'm down to about 15 percent on manual.
04 41 58	8.9	CC	Roger. You're going to use fly-by-wire for reentry and we recommend that you do the best you can to keep a zero angle during reentry. Over.
04 42 07	1.2	P	Roger. Friendship Seven.
04 42 11	3.4	P	This is Friendship Seven. I'm on fly-by-wire, back it up with manual. Over.
04 42 16	1.1	CC	Roger, understand.
04 42 27	9.2	CC	Seven, this is Cape. The weather in the recovery area is excellent, 3-foot waves, only one-tenth cloud coverage, 10 miles visibility.
04 42 37	1.2	P	Roger. Friendship Seven.
04 42 45	1.4	CC	Seven, this is Cape. Over.
04 42 47	2.7	P	Go ahead, Cape, you're ground, you are going out,
04 42 50	1.8	CC	We recommend that you ....
04 43 14	2.9	P	This is Friendship Seven. I think the pack just let go.
04 43 37	2.4	P	This is Friendship Seven. A real fireball outside.
04 44 18	1.9	P	Hello, Cape. Friendship Seven. Over.
04 45 16	1.9	P	Hello, Cape. Friendship Seven. Over.
04 45 41	2.3	P	Hello, Cape. Friendship Seven. Do you receive? Over.
04 46 17	2.0	P	Hello, Cape. Friendship Seven. Do you receive? Over.
04 47 15	1.2	CC	... How do you read? Over.
04 47 16	1.5	P	Loud and clear; how me?
04 47 19	1.6	CC	Roger, reading you loud and clear. How are you doing?
04 47 22	1.0	P	Oh, pretty good.

04 47 26	3.8	CC	Roger. Your impact point is within I mile of the up-range destroyer.
04 47 30	0.5	P	Roger.
04 47 31	0.2	CC	... Over.
04 47 32	0.3	P	Roger.
04 47 40	3.4	CC	This is Cape, estimating 04 50. Over.
04 47 44	1.5	P	Roger, 04 50.
04 47 49	1.6	P	Okay, we're through the peak g now.
04 47 51	4.0	CC	Seven, this is Cape. What's your general condition? Are you feeling pretty well?
04 47 55	2.8	P	My condition is good, but that was a real fireball, boy.
04 48 01	3.2	P	I had great chunks of that retropack breaking off half the way through.
04 48 04	2.1	CC	Very good; it did break off, is that correct?
04 48 07	3.4	p	Roger. Altimeter off the peg indicating 80,000.
04 48 11	1.7	CC	Roger, reading you loud and clear.
04 48 13	0.3	P	Roger
04 48 19	6.7	CC	Seven, this is Cape. You're ... will be within I mile of the up-range destroyer. Recovery weather is very good. Over.
04 48 26	5.0	P	Roger, understand. 55,000, standby, MARK.
04 48 37	3.3	P	I'm getting all kinds of contrails and stuff outside out here.
04 48 42	2.7	CC	Roger. Say again your altitude, please. You were broken up.
04 48 45	0.3	P	45,000.
04 48 51	3.7	P	Rocking quite a bit. I may still have some of that pack on. I can't damp it either.
04 49 00	2.0	CC	... post reentry check list. Over.
04 49 12	4.2	P	Friendship Seven. Going to drogue early. Rocking fairly, drogue came out.
04 49 18	0.8	P	Drogue is out.
04 49 20	4.0	P	Roger, drogue came out at 30,000, at about a 90° yaw.
04 49 25	1.6	CC	Roger, is the drogue holding all right?
04 49 27	1.1	P	Roger, the drogue looks good.
04 49 29	0.4	CC	Roger.
04 49 31	1.6	P	Scope did not come out.
04 49 32	1.0	CC	... check list.
04 49 34	1.5	P	Roger, pumping the scope out.
04 49 36	1.2	CC	... check list. Over.
04 49 38	0.5	P	Say again.
04 49 42	3.6	P	Roger, reentry checklist complete. Standing by for main at ten [thousand feet].
04 49 47	0.4	CC	Roger.
04 50 00	2.2	P	Coming down on ten [thousand feet], snorkels are open.
04 50 04	1.8	CC	Roger, understand snorkels open.
04 50 07	0.5	P	Roger.
04 50 10	20.8	P	Main chute in on green. Chute is out, in reef condition at 10,800 feet and beautiful chute. Chute looks good. On O2 emergency and the chute looks very good. Rate of descent has gone to about 42 feet per second. The chute looks very good.

CANAVERAL-RECOVERY

04 50 37	3.4	P	Hello, Mercury Recovery. This is Friendship Seven. Do you receive?
04 50 41	4.2	R	Mercury Friendship Seven, this Steelhead. Loud and clear. Over.
04 50 45	4.1	P	Roger, Steelhead. Friendship Seven. The chute looks very good. Over.
04 50 51	4.3	R	Roger, understand the chute very good, descent normal. Is that Charlie? Over.
04 50 55	5.2	P	Roger, that is affirmative. Descent is normal, indicating 40 feet per second.
04 51 05	2.9	P	My condition is good; it's a little hot in here, however. Over.

04 51 09	8.0	R	Roger, Friendship Seven. Be advised, I got your chaff on my radar and I'm heading out for you now. Over.
04 51 17	2.8	P	Roger. What is your estimate on recovery time? Over.
04 51 23	2.0	R	Friendship Seven. Steelhead. Wait. Out.
04 51 26	2.6	P	Roger. Friendship Seven. Indicating 7,000.
04 51 35	7.8	R	Friendship Seven, this is Steelhead. Be advised, I have you visually, estimate on station in approximately 1 hour. Over.
04 51 43	1.8	P	Roger, on station 1 hour.
04 51 47	2.2	P	This is Friendship Seven, standing by for impact.
04 51 54	1.8	P	This is Friendship Seven, going through checklist.
04 52 04	6.6	R	Friendship Seven, this is Steelhead. Correct my estimate on station. Estimate pickup now at 20 minutes. Over.
04 52 10	2.5	P	Roger, understand 20 minutes to pickup.
04 52 14	2.0	p	Friendship Seven, going through check list.
04 52 27	3.0	CC	Seven, this is Cape. Do you have the Landing Bag on green? Over.
04 52 32	1.6	p	Friendship Seven- Say again, Cape.
04 52 35	1.4	CC	. . . Landing Bag on green? Over.
04 52 38	2.9	P	This is Friendship Seven. Still did not receive you. Repeat again.
04 52 43	3.1	CC	Do you have Landing Bag on green? Over.
04 52 47	2.8	p	I'm sorry. I cannot read you, Cape. Say again.
04 52 51	3.4	R	Do you have Landing Bag on green? Over.
04 52 55	2.6	p	That's affirmative. Landing Bag is on green.
04 53 04	6.9	R	Friendship Seven, this is Steelhead. Be advised accordingly to my surface gadget, your range 6 miles from me, on the way. Over.
04 53 11	2.4	P	Roger, understand 6 miles. Good show.
04 53 57	0.3	P	Okay.
04 53 58	2.1	CC	Steelhead, this is Cape Cap Com. Over.
04 54 05	1.0	P	Go ahead, Cap Com.
04 54 07	7.3	CC	Cape Cap Com. We recommend that he remain in the capsule unless he has some overriding reason for getting out. Over.
04 54 17	2.1	P	Say again. This is Friendship Seven.
04 54 22	5.0	R	Remain in capsule unless you have an overriding reason for getting out. Over.
04 54 28	1.5	P	Roger. Friendship Seven.
04 54 32	2.3	P	Friendship Seven. Ready for impact; almost down.
04 54 35	0.7	R	Roger.
04 54 37	1.1	P	Do you have me in sight?
04 54 47	2.2	P	Friendship Seven. Getting close. Standing by.
04 54 50	0.6	R	Roger.
04 55 10	1.0	P	Here we go.
04 55 20	2.6	P	Friendship Seven. Impact. Rescue Aids is manual.

RECOVERY

04 55 47	4.3	R	Friendship Seven, this is Steelhead. Hold you in the water. What is your condition? Over.
04 55 51	3.4	P	Roger, my condition okay. Does the capsule look like it's okay? Over.
04 55 56	4.0	R	Friendship Seven, reference your last; affirmative, capsule looks good from here. Over.
04 56 01	3.8	P	Roger, understand they want me to stay in the capsule until rescue.
04 56 06	2.1	R	Friendship Seven, Steelhead. That's Charlie. Over.
04 56 09	0.6	P	Roger.
04 57 00	1.8	R	Friendship Seven, Steelhead is calling you.
04 57 03	2.6	P	Go ahead, Steelhead. Friendship Seven. I don't receive you.
04 57 11	2.3	R	Go ahead, Steelhead. Friendship Seven reads you.
04 57 16	2.7	P	Negative. This is Friendship Seven. I do not read him.

04 57 19	2.9	R	Friendship Seven, Friendship Seven, this is Steelhead, Steelhead. Over.
04 57 23	1.7	P	Go ahead, Steelhead. Friendship Seven.
04 57 25	6.2	R	Friendship Seven, this is Steelhead. I understand you have me visually through your window. Is that affirmative? Over.
04 57 32	0.6	P	Negative.
04 57 36	1.4	P	Negative. This is Friendship Seven.
04 57 55	7.4	R	Friendship Seven, this is Steelhead. Estimate recovery time in approximately 7 minutes. Over.
04 58 03	4.2	P	Roger, 7 minutes. Understand you're going to put men in the water with the collar. Is that affirm?
04 58 09	2.1	R	Friendship Seven, Steelhead. Affirmative. Over.
04 58 12	0.6	P	Roger.
04 58 31	5.9	R	Friendship Seven, this is Steelhead. Understand your condition excellent at this time. Is that Charlie?
04 58 37	5.4	P	That's affirmative. My condition is good. I'm a little warm at the moment, but that's okay, the suit fans are still running.
04 58 44	1.2	R	Steelhead. Roger. Out.
04 59 38	1.7	CC	Steelhead, this is Cape. Over.
04 59 41	4.2	R	Station calling . . . . Read you. Say again. Over.
04 59 46	2.9	CC	This is Cape Canaveral. Recommend . . . . Over.
04 59 52	6.4	R	Understand station calling Steelhead is the Cape, unable to read your message. Request you say again text. Over.
04 59 58	5.0	CC	Roger. Recommend . . . astronaut. Over.
05 00 08	5.5	R	Cape, this is Steelhead; request you say again all after recommend. Over.
05 00 16	9.6	CC	Keep the Astronaut advised of the recovery progress; keep the Astronaut advised of the recovery progress. Over.
05 00 27	1.4	R	Steelhead. Wilco. Out.
05 01 02	9.2	R	Friendship Seven, this is Steelhead. We have you visually. I am closing now. Should be, should be ready to effect recovery in approximately 4 minutes. Over.
05 01 12	3.0	P	Roger, 4 minutes to recovery and my condition is good.
05 01 17	15.0	R	Steelhead. Roger. Break, have ground tackle, etc., standing by on deck. All equipment fully rigged. Believe we will be able to have you aboard in approximately 2 minutes after arrival. Over.
05 01 32	0.5	P	Roger.
05 01 38	9.4	P	This is Friendship Seven. I'm very warm. I'm not, I'm just remaining, remaining motionless here trying to keep as cool as possible. I'm extremely warm at the moment.
05 01 49	5.9	R	Steelhead. Roger, break, disregard that.
05 01 55	0.3	P	Roger.
05 02 11	7.1	R	Friendship Seven, Steelhead. Helicopter is on its way. At present, they expect to be here in approximately 8 minutes. Over.
05 02 18	0.6	P	Roger.
05 02 29	8.7	R	Friendship Seven, Steelhead. Medico are standing by in case assistance necessary immediately after recovery. Over.
05 02 39	1.3	P	Roger. Friendship Seven.
05 03 04	4.6	R	Friendship Seven, Steelhead. My speed now 10 [knots), commencing my approach. Over.
05 03 09	1.1	P	Roger. Friendship Seven.
05 03 18	6.0	R	Friendship Seven, this is Steelhead. All equipment operating normally. Expect to be along side approximately 3 minutes. Over.
05 03 26	0.5	P	Ah, Roger.
05 03 57	8.6	R	Friendship Seven, this is Steelhead. All stop, I say again, my engines are stop. I'm coming along side at this time. Over.
05 04 05	0.6	P	Roger.

05 04 09	2.8	R	Friendship Seven, Friendship Seven, this is Steelhead. Do you copy? Over.
05 04 12	4.0	P	Roger, Steelhead, I copy. Friendship Seven. Understand you're coming along side.
05 04 28	14	R	Friendship Seven, this is Steelhead. You are now 1,000 yards. Over.
05 04 33	3.9	P	Roger. Friendship Seven. Sounds good.
05 04 40	3.8	P	Capsule looks to me as though its floating in pretty good shape. Does it look that way to you?
05 04 46	7.6	R	Friendship Seven, this is Steelhead. Affirmative, capsule looks good from here. I can, can discern no damage visually. Over.
05 04 54	0.6	P	Roger.
05 04 58	6.3 (Aircraft)	R	Friendship Seven, this is 6 Spangle 8. I'm orbiting at 300 feet, everything looks perfectly normal from here.
05 05 05	0.5	P	Roger.
05 05 08	2.0 (Aircraft)	R	You're riding in a good attitude. Over.
05 05 10	0.3	P	Roger.

APPENDIX C

DESCRIPTION OF THE MA-6 ASTRONOMICAL, METEOROLOGICAL, AND TERRESTRIAL OBSERVATIONS.

By JOHN H. GLENN, Jr., Astronaut, NASA Manned Spacecraft Center
This material is taken verbatim from the transcript Of Pilot's Postflight debriefing on Grand Turk island, Feb. 21, 1962.

Luminous Particles

Coming out of the night on the first orbit, at the first glint of sunlight on the capsule, I was looking inside the capsule to check some instruments for probably 15 or 20 seconds. When I glanced back out the window, my initial reaction was that the capsule (spacecraft) had tumbled and that I was looking off into a star field and was not able to see the horizon. I could see nothing but luminous specks about the size of the stars outside. I realized, however, they were not stars. I was still in the attitude that I had before. The specks were luminous particles that were all around the capsule. There was a large field of spots that were about the color of a very bright firefly, a light yellowish green color. They appeared to vary in size from maybe just pinhead size up to possibly 3/8 of an inch. I would say that most of the particles were similar to first magnitude stars; they were pretty bright, very luminous. However, they varied in size so there would be varying magnitudes represented. They were floating in space at approximately my speed. I appeared to be moving through them very slowly, at a speed of maybe 3 to 5 miles an hour. They did not center on the capsule as though the capsule was their origin. I thought first of the lost Air Force needles that are some place in space but they were not anything that looked like that at all.

The other possibility that came to my mind immediately was that snow or little frozen water particles were being created from the peroxide decomposition. I don't believe that's what it was, however, because the particles through which I was moving were evenly distributed and not more dense closer to the capsule.

As I looked out to the side of the capsule, the density of the field to the side of the capsule appeared to be about the same as directly behind the capsule. The distance between these particles would average, I would estimate, some 8 or 10 feet apart. Occasionally, one or two of them would come swirling up around the capsule and across the window, drifting very, very slowly, and then would gradually move off back in the direction I was looking. This was surprising, too, because it showed we probably did have a very small flow field set up around the capsule or they would not have changed their direction of motion as they did. No, I do not recall observing any vertical or lateral motion other than that of the particles that swirled around close to the spacecraft. It appeared to me that I was moving straight through a cloud of them at a very slow speed. I observed these luminous objects for approximately 4 minutes before the sun came up to a position where it was sufficiently above the horizon that all the background area then was lighted and I no longer could see them.

After passing out of them, I described them as best I could on the tape recorder and reported them to the Cape. I had two more chances to observe them at each sunrise; it was exactly the same each time. At the first rays of the sun above the horizon, the particles would appear. To get better observation of these particles and to make sure they were not emanating from the capsule, I turned the capsule around during the second sunrise. When I turned around towards the sunrise, I could see only 10 percent as many particles as I could see when facing back toward the west. Still, I could see a few of them coming toward me. This proved rather conclusively, to me at least, that I was moving through a field of something and that these things were not emanating, at least not at that moment, from the capsule. To check whether this might be snowflakes from the condensation from the thrusters, I intentionally blipped the thrusters to see if I was making a pattern of these particles. I could observe steam coming out of the pitchdown thruster in good shape and this didn't result in any observation of anything that looked like the particles. I had three good looks at them and they appeared identical each time. I think the density of the particles was identical on all three passes.

I would estimate that there were thousands of them. It was similar to looking out across a field on a very dark night and seeing thousands of fireflies. Unlike fireflies, however, they had a steady glow. Once in a while, one or two of them would come drifting up around the corner of the capsule and change course right in front of me. I think it was from flow of some kind or perhaps the particles were ionized and were being attracted or repelled. It was not due to collisions because I saw some of them change course right in front of me without colliding with any other particles or the spacecraft. If any particles got in near enough to the capsule and got into the shade, they seemed to lose their luminous quality. And when occasionally I would see one up very close, it looked white, like a little cottony piece of something, or like a snowflake. That's about the only description of them I have. There was no doubt about their being there because I observed them three different times for an extended period of time. I tried to get pictures of them, but it looks like there wasn't sufficient light emanating from them to register on the color film.

The High Layer

I had no trouble seeing the horizon on the nightside. Above the horizon some 6 to 8 degrees, there was a layer that I would estimate to be roughly 1½ to 2 degrees wide. I first noticed it as I was watching stars going down. I noticed that as they came down close to the horizon, they became relatively dim for a few seconds, then brightened up again and then went out of sight below the horizon. As I looked more carefully, I could see a band, parallel to the horizon, that was a different color than the clouds below. It was not the same white color as moonlight on clouds at night. It was a tannish color or buff white in comparison to the clouds and not very bright. This band went clear across the horizon. I observed this layer on all three passes through the nightside. The intensity was reasonably constant through the night. It was more visible when the moon was up but during that short period when the moon was not up, I could still see this layer very dimly. I wouldn't say for sure that you could actually observe the specific layer during that time, but you could see the dimming of the stars. But, when the moon was up, you very definitely could see the layer, though it did not have sharp edges. It looked like a dim haze layer such as I have seen occasionally while flying. As stars would move into this layer, they would gradually dim; dim to a maximum near the center and gradually brighten up as they came out of it. So, there was a gradient as they moved through it; it was not a sharp discontinuity.

Nightside Observations of the Earth

Over Australia, they had the lights of Perth on and I could see them well. It was like flying at high altitude at night over a small town. The Perth area was spread out and was very visible and then there was a smaller area south of Perth that had a smaller group of lights but they were much brighter in intensity; very luminous. Inland, there were a series of about 4 or 5 towns that you could see in a row lined up pretty much east and west that were very visible. It was very clear; there was no cloud cover in that area at that time.

Knowing where Perth was, I traced a very slight demarcation between the land and the sea, but that's the only time I observed a coastline on the nightside. Over the area around Woomera, there was nothing but clouds. I saw nothing but clouds at night from there clear up across the Pacific until we got up east of Hawaii. There was solid cloud cover all the way.

In the bright moonlight, you could see vertical development at night. Most of the areas looked like big sheets of stratus clouds, but you could tell where there were areas of vertical development by the shadows or lighter and darker areas on the clouds.

Out in that area at night, fronts could not be defined. You can see frontal patterns on the dayside. In the North Atlantic, you could see streams of clouds, pick out frontal areas pretty much like the pictures from earlier Mercury flights.

With the moonlight, you are able to pick up a good drift indication using the clouds. However, I don't think it's as accurate as the drift indications during the day. The drift indication is sufficient that you can at least tell what direction you're going at night within about 10 or 15 degrees. In the daylight over the same type clouds, you probably could pick up your drift down to maybe a couple of degrees.

The horizon was dark before the moon would come up, which wasn't very long. However, you can see the horizon silhouetted against the stars. It can be seen very clearly. After the moon comes up, there is enough light shining on the clouds that the earth is whiter than the dark background of space. Well, before the moon comes up, looking down is just like looking into the black hole at Calcutta.

There were a couple of large storms in the Indian Ocean. The Weather Bureau scientists were interested in whether lightning could be seen or not. This is no problem; you can see lightning zipping around in these storms all over the place. There was a great big storm north of track over the Indian Ocean; there was a smaller one just south of track and you could see lightning flashing in both of them; especially in the one in the north; it was very active. It was flashing around and you could see a cell going and another cell going and then horizontal lightning back and forth.

On that area, I got out the airglow filter and tried it. I could not see anything through it. This, however, may have been because I was not well enough dark adapted. This is a problem. If we're going to make observations like this, we're going to have to figure out some way to get better night adapted in advance of the time when we want to make observations. There just was not sufficient time. By the time I got well night adapted, we were coming back to daylight again.

Dayside Observations

Clouds can be seen very clearly on the daylight side. You can see the different types, vertical developments, stratus clouds, little puffy cumulus clouds, and alto-cumulus clouds. There is no problem identifying cloud types. You're quite a distance away from them, so you're probably not doing it as accurately as you could looking up from the ground, but you can certainly identify the different types and see the weather patterns. The cloud area covered most of the area up across Mexico with high Cirrus almost to New Orleans. I could see New Orleans; Charleston and Savannah were also visible.

You can see cities the size of Savannah and Charleston very clearly. I think the best view I had of any area during the flight was the clear desert region around El Paso on the second pass. There were clouds north of Charleston and Savannah, so I could not see the Norfolk area and on farther north. I did not see the Dallas area that we had planned to observe because it was covered by clouds but at El Paso, I could see the colors of the desert and the irrigated areas north of El Paso. You can see the pattern of the irrigated areas much better than I had thought we would be able to. I don't think that I could see the smallest irrigated areas; it's probably the ones that are blocked in by the larger sized irrigation districts which I saw. You can see the very definite square pattern in those irrigated areas, both around El Paso and at El Centro which I observed after retrofire.

The western part of Africa was clear. That is, a desert region where I mainly saw dust storms. By the time we got to the region where I might have been able to see cities in Africa, the land was covered by clouds. I was surprised at what a large percentage of the track was covered by clouds on this particular day. There was very little land area which could be observed on the daylight side. The eastern part of the United States and occasional glimpse of land up across Mexico and the desert area in Western Africa was all that could be seen. I saw what I assume was the Gulf Stream. The water can be seen to have different colors. Another thing that I observed was the wake of a ship as I came over Recovery Area G at the beginning of the third orbit. I had pitched down to below retro-attitude. I was not really thinking about looking for a ship. I was looking down at the water and I saw a little V. I quickly broke out the chart and checked my position. I was right at Area G, the time checked out perfectly for that area. So, I think I probably saw the wake from a recovery ship when I looked back out and tried to locate it again and the little V had gone under a cloud and I didn't see it again. The little V was heading west at that time. It would be interesting to see if the carrier in Area G was fired up and heading west at that time.

I would have liked to put the glasses on and see what I could have picked out on the ground. Without the glasses, I think you identify the smaller objects by their surroundings. For instance, you see the outline of a valley where there are farms and the pattern of the valley and its rivers and perhaps a town. You can see something that crosses a river and you just assume that it's a bridge. As far as being able to look down and see it and say that is a bridge, I think you are only assuming that it's a bridge more than really observing it. Ground colors show up just like they do from a high-altitude airplane; there's no difference. A lot of the things you can identify just as from a high flying airplane. You see by color variations the deep green woods and the lighter green fields and the cloud areas.

I could see Cape Canaveral clearly and I took a picture which shows the whole Florida Peninsula; you see across the interior of the Gulf.

Sunset and Sunrise Horizon Observations

At sunset, the flattening of the sun was not as pronounced as I thought it might be. The sun was perfectly round as it approached the horizon. It retained its symmetry all the way down until just the last sliver of sun was visible. The horizon on each side of the sun is extremely bright and when the sun got down to where it was just the same level as the bright horizon, it apparently spread out perhaps as much as 10 degrees each side of the area you were looking. Perhaps it was just that there was already a bright area there and the roundness that had been sticking up above it came down to where finally that last little sliver just matched the bright horizon area and probably added some to it.

I did not see the sunrise direct; only through the periscope. You cannot see that much through the scope. The sun comes up so small in the scope that all you see is the first shaft of light. The band of light at the horizon looks the same at sunrise as at sunset.

The white line of the horizon is extremely bright as the sun sets, of course. The color is very much like the arc lights they use around the pad.

As the sun goes on down a little bit more, the bottom layer becomes a bright orange and it fades into red; then on into the darker colors and finally off into blues and black as you get further up towards space. One thing that was very surprising to me, though, was how far out on the horizon each side of that area the light extends. The lighted area must go out some 60 degrees. I think this is confirmed by the pictures I took.

I think you can probably see a little more of this sunset band with the eye than with a camera. I was surprised when I looked at the pictures to see how narrow looking it is. I think you probably can pick up a little broader band of light with the eye than you do with the camera. Maybe we need more sensitive color film.

APPENDIX D

PRELIMINARY REPORT ON THE RESULTS OF THE MA-6 FLIGHT IN THE FIELD OF SPACE SCIENCE

By JOHN A. O'KEEFE, Ph. D.; Asst. Chief, Theoretical Division, NASA Goddard Space Flight Center

Introduction

This paper discusses the preliminary attempts to explain the observations made by Astronaut Glenn during the MA-6 flight. Analysis of Pilot Glenn's observations is continuing and is not yet complete. This paper is intended only to indicate the direction which the analysis is taking, not to provide the final explanations. The theories presented are those of the author, not the astronaut. In some cases, final verification of these theories must await further Mercury flights.

Four principal points are to be considered in the field of space science as a result of the MA-6 flight. They are:

(1) The luminous particles (Glenn effect), which are probably the result of the flaking off of paint, or possibly the condensation of moisture from the spacecraft heat exchanger

(2) A luminous band seen around the sky and possibly due to airglow or aurora but probably due to reflections of the horizon between the windows of the spacecraft

(3) The flattened appearance of the sun at sunset. This is not attested by the visual observations, but appears fairly clear in the photographs

(4) The ultraviolet photography.

Luminous Particles

Glenn observed a field of small, luminous objects surrounding his spacecraft at sunrise on all three orbits. He compares them to fireflies, especially in color, remarking that they were very luminous and variable in size. Some of these particles came close to the spacecraft so that they got into the shade, as witnessed by a marked loss in brightness, and a change in color from yellow-green to white. The change in color is comprehensible as being due to passage from illumination by direct sunlight to illumination by bluish light scattered from the twilight all along the horizon. Passage into the shadow is a clear indication that the particles involved were genuinely close at hand. It indicates that the particles were within the range of stereoscopic vision, so that Glenn's distance estimates are meaningful. It follows that his estimates of relative velocity are also meaningful. These estimates were 3 to 5 miles per hour, that is, 1.3 to 2.2 meters per second relative to the spacecraft. Glenn stated that the overall impression was that the spacecraft was moving through a field of these particles at the above speed.

Evidence That Particles Are Associated With Spacecraft

This observation indicates that the luminous objects were undoubtedly associated with the spacecraft in their motion. The spacecraft velocity was approximately 8,000 meters per second; the velocity of the particles was identical with that of the spacecraft in all three coordinates within about 1 part in 4,000. Rough estimates show that this implies that the orbital inclination was the same for the particles as for the spacecraft within $\pm 0.01°$. The eccentricity was the same within ± 0.0002. In particular, the spacecraft was at that time descending toward perigee at the rate of approximately 50 meters per second. The particles were descending at the same rate within ± 2 meters per second. Thus, from velocity consideration alone, there is a very convincing demonstration that the particles were associated with the spacecraft.

In addition, it should be noted that the height at that time was 160 kilometers. It was thus at least twice the height of the noctilucent clouds (which apparently consist of ice particles, and must therefore be considered). At this level, the atmosphere has a density of the order of 10^{-10} grams/cm^3; it is completely unable to retard the fall of any visible object. Hence, there is no reason to expect any layer of particles sustained at this level. Anything at this height must be in orbit.

Evidence That the Size of the Field of Particles Must Be Relatively Small

An important consideration is the fact that the field of particles could not have been of very great extent. If, for example, we suppose that there were two or three of these "very luminous" particles within 3 meters of the window (the spacing being estimated by Glenn at 6 to 10 feet, or 2 to 3 meters), then in the next 3 meters, there should be 12 particles, averaging one fourth as bright so that the contribution to the total illumination from the second 3 meters is the same as from the first 3, and so on. Had the field extended to a distant of "several miles," that is, say 10 kilometers, the total light would have been some 3,000 times that of the individual nearby particles, and Glenn would have spoken of an intensely luminous fog. Since he saw this for a time of about 4 minutes, during which he traveled about 1,920 kilometers, the field, if a part of the environment, would have been of this length and the particles would have covered the sky solidly in this direction, so that it would have looked like a cloud or a snowfield. This sort of calculation is well known in astronomy under the name of Olber's paradox. It establishes with certainty that the particles did not extend far in any direction from the spacecraft. The fact that Glenn did not see a local concentration around the spacecraft means that there was no large density increase within the range of stereoscopic vision, but it does not conflict with the idea that the field extended at most, a few hundred meters in any direction.

Evidence for Particle Size and Brightness

With respect to the brightness of the particles, conversations with Astronaut Glenn have established that the most significant brightness estimate is the comparison with fireflies. Mr. T. J. Spilman, of the Smithsonian Institution, states that the available measures of light of Photinus pyralis, the common firefly of the eastern United States, indicate from 1/50 to 1/400 candle, when the light is turned on. At a distance of 1 meter, a candle has a brightness of about - 14; the firefly at 2 meters would be 200 to 1,600 times fainter, or between about -8.3 and - 6. At distances of the order of 20 meters, it would be from - 3.3 to - 1, and thus comparable with planets or the brightest stars.

The full moon (- 12.6) is plainly visible on several of the photographs taken in orbits. The particles may possibly also be visible; but if so, they are not more than 1/10 the brightness of the full moon, and hence not brighter than about - 10. Of course occasionally a large particle may have come close; but the run of the mine must have been - 10 or fainter.

A white object 1 centimeter in diameter, at a distance of 2 meters in direct sunlight would be about - 19.9 magnitude; if of pinhead size (2 millimeters) it would be - 10.4. If we allow a reduction to 1 millimeter on account of the known effect of bright objects to seem larger than they are, we find - 9, which is of the same order as the firefly at the same distance.

Probable Cause of Particle Motion

The next question is, what is the agency which is causing the particles to move with respect to the spacecraft? The possibilities are electrical, magnetic, and gravitational fields; light pressure; and aerodynamic drag. Of these, the electrical forces can be discarded for mass motion over a large area, since we are in the lower F region of the ionosphere and space is essentially a conductor. Magnetic fields can be divided into terrestrial and spacecraft. The spacecraft field is certainly too small at reasonable distances, and the terrestrial field cannot accelerate a dipole, because the field gradient is too small. Gravitational fields will act in almost precisely the same way on the spacecraft as on the particles. The acceleration difference will be one way for those below the spacecraft and the other way for those above; thus, they will make the particles seem to go around the spacecraft with a steady motion, rather than to move past it.

Light pressure and drag have similar effects at sunrise, but at heights of the order of 160 kilometers, drag is about 1 dyne per square centimeter, while radiation pressure is less by many orders of magnitude. Hence, the most probable source of the acceleration is aerodynamic drag.

Nature of Particles

Important information about the nature of the particles is furnished by their behavior under the influence of drag forces. At sunrise, the spacecraft was a little above its minimum altitude of 160 kilometers. At this height, the density of the air is roughly 1.3×10^{-12} gm/ cm^3; the spacecraft velocity is about 8×10^5 cm/sec; the drag pressure is thus about 1 dyne/cm^2. Since Glenn states that he appeared to be moving slowly through a relatively stationary group of particles, it is evident that they could not have been greatly accelerated while in the near vicinity of the spacecraft. For comparison, a snowflake with a diameter of 1 millimeter and the usual density of 0.1 will have about 0.01 gram per square centimeter of frontal area. It will thus be accelerated at the rate of 100 cm/sec^2, and will exceed the estimated velocities after only 2 seconds, when it has gone 2 meters. We cannot escape from the problem by supposing the snowflakes to be much larger, say 1 centimeter in diameter because, though occasional particles may have been as large as this, the majority must have been smaller because they did not give strong photographic images.

Glenn tells us that their average separation was only about 6 to 10 feet, so that at any moment one would be expected to be within a few meters of the spacecraft window, and hence brighter than the full moon.

A few particles, which came close to the window and could be examined in detail appeared large and cottony. These were very likely snowflakes. They were seen to accelerate perceptibly in the air stream.

We are now in a position to attempt to estimate the material of the particles. It is clear at once that we are not dealing with any sort of gas fluorescence or gas discharge, such as might be produced by the motion of the spacecraft through the ionosphere, because the lights were not visible until sunrise. They were, therefore, shining by reflected light. Solid or liquid particles are more efficient in reflecting light than gases by factors of millions; hence the particles must be taken as solid or liquid. Their sizes were probably in the millimeter range, as judged from their apparent brightness. Their densities must have been much higher than 0.1. The highest densities reasonable may be about 3; in this case the particles would be accelerated at 3 cm/sec^2, and would reach a velocity of 2 meters/sec after a time of 1 minute when they would be 50 meters away, and their velocities would be difficult to estimate accurately.

Particles Did Not Originate From Launch Vehicle

It can be shown at this point that the particles could not have come from the sustainer, which was over 100 kilometers away at the first sighting, and about 300 kilometers away at the third sighting. If accelerated over this distance at the lowest reasonable rate, namely 3 cm/sec^2, they would have passed the spacecraft at 135 meters per second, which cannot be reconciled with the observations. Any small particles observed at this altitude moving with low relative velocity must have been released from the spacecraft itself, and not very long previously.

Another significant item is the total mass. With a separation of the order of 3 meters (10 feet) as reported by Glenn, there is about 1 particle per 30 cubic meters; the particles apparently weigh about 3 milligrams each. If a 100-meter cube is imagined filled with these particles, there will be about 30,000, with a total weight of about 1 kilogram. If we assume them to be 1 centimeter snowflakes, of mass 100 milligrams each, the total weight is 30 kilograms. Since Glenn reports the field as extending widely, it is clear that the denser smaller particles are more probable.

Possible Sources of Particles in Spacecraft

Among the materials known to have come off the spacecraft, only the three following appear to have had sufficient volume:

(1) Solid particles: A considerable amount of paint flaked off the outside of the spacecraft; in addition, it is possible that some solid particles were flaked off paint and other materials in the area between the heat shield and the pressure vessel

(2) Water from the hydrogen peroxide thrusters

(3) Water from the cooling system.

Among these, (2) can be discarded at once, first, because Glenn himself directly studied this possibility in flight by watching the output of the pitch-down thruster. He noted at that time that the jet of steam, which was visible, was entirely unlike the observed particles. In the second place, the velocity imparted to the steam as a necessary part of the thruster operation would have taken the steam away immediately.

The water from the cooling system may well have been responsible for a few large snowflakes which Glenn described. This water, after being used to cool the spacecraft, is released through a hole, about 2.5 centimeters across, into the space between the spacecraft bulkhead and the heat shield. This space is approximately 10 centimeters in depth and extends over the back of the heat shield, which is about 2 meters in diameter. The volume is thus roughly 3 X 105 cubic centimeters, or 300 liters. From this space, it emerges through 10 or more holes, each about 1 centimeter in diameter, spaced around the heat shield.

This system appears likely to produce snowflakes. When operated in tests, the clogging of the 2.5 centimeter pipe by ice was a common occurrence. In flight this condition was also indicated by warning lights, on the MA-6 flight. Vapor which got through the 2.5 centimeter pipe to the space back of the bulkhead would expand against the low pressure inside the bulkhead and would cool. Ice crystals would form, but these might not leave the spacecraft for some time, because of the smallness of the ports relative to the size of the space. This situation, where a low gas pressure might be sustained for a considerable period, is very helpful in understanding the growth of snowflakes as large as 1 centimeter in diameter. It would be hard to see how such flakes could grow in empty space.

As a result of the relatively low temperatures, the large size of the pipes, and the cooling and the condensation back of the bulkhead, the gas pressure at the ports would be expected to be very low, so that the snowflakes would emerge with low velocities, as described by Glenn. It is easy to imagine a flake formed in this way drifting down past the spacecraft window slowly in the manner described. As long as it was back of the heat shield, it would not experience the air stream; but eventually, as described by Glenn, it would drift up into the air stream and then start moving up to the rear. Such particles would look like white cottony snowflakes because they were. Their color would be different by direct sunlight from their color in the shadow for the same reason as the phenomenon that shadows at sunset are sometimes blue. (See ref. 1.) The light that gets into the shadow is the light from the long twilight arc on the earth, and this is predominantly blue.

The total quantity of water available from this source is about 1 kilogram per hour. In view of the very short time that it could remain in the vicinity of the spacecraft and the relatively large total amount required to fill a reasonable volume around the spacecraft, it appears somewhat unlikely that ice is the material of the particles, though the possibility cannot be entirely excluded that dense ice crystals were involved.

Another possibility is solid particles of material such as paint. Millimeter size particles of this type would have densities on the order of 3 and masses of the order of 3 milligrams. Within a sphere of radius 10 meters around the spacecraft, there would be 140 such particles, with a total mass of about ½ gram. Within 100 meters, there would be about ½ kilogram. If we suppose that particles of this type are liberated primarily at sunrise, possibly because of some cracking or stretching of the spacecraft skin occurring at this time, it is not necessary to imagine much more than 1½ kilograms of material was liberated during the whole flight, especially if we suppose that the density was somewhat less in the outer portions of the cloud. This figure is perhaps not inconsistent with the amount of material which could have flaked off. It is necessary to emphasize the extremely tenuous character of these figures which depend on estimates of the cloud size, since the mass of material required varies with the cube of the diameter of the cloud.

Summing up, it appears that the Glenn effect is due to small solid particles, mostly about 1 millimeter in diameter, but with a few larger bodies in addition. The brightness of the majority of the particles was about -9 at a distance of 2 meters. They were probably at least as dense as water; higher densities are more likely. They were certainly not a part of the space environment, but were something put in orbit as a result of the MA-6 flight. They were almost as certainly related to the spacecraft, not to the sustainer. Two reasonable possibilities exist, namely ice from the cooling system, and/or particles of paint or other heavy material which flaked off the spacecraft under the low pressures of the space environment. Of these, the paint is the more probable because its higher density explains the orbital behavior better, and because we can understand why paint might be liberated only at sunrise, while ice would be liberated throughout the flight. Hence, very large quantities of ice would be needed, compared to the amounts of paint. In short, the most probable explanation of the Glenn effect is millimeter-size flakes of material liberated at or near sunrise by the spacecraft.

The Luminous Band

On all three revolutions, Glenn reports a luminous band, at a height of 7° to 8° above the horizon, tan to buff in color, more luminous when the moon (then full) was up. The band is stated to have been faintly and uncertainly visible when the moon was down; at that time, the horizon was clearly seen silhouetted by stars. After the flight, it was noted that many photographs of the twilight showed a luminous band parallel to the horizon. Photographs of the sky in full daylight showed a faint luminous zone extending all the way up from the horizon. The faintness of the band on daylight photographs was probably due to the automatic reduction of the exposure in strong light.

The focal length of the camera lens was 50 millimeters. Then photographs were enlarged about 6.8 times for study, yielding a scale about 0°.17 per millimeter. The height of the band seen on the enlargements was about 75 millimeters, or 12°.6.

The band seen on the photographs had not been noted as such in the spacecraft. It was therefore thought at first to be perhaps a camera effect. However, the circular symmetry of a camera lens makes it difficult to explain a band parallel to the horizon.

The most probable explanation of the luminous band seen on the photographs is multiple reflections within the spacecraft window system. The spacecraft has an inner and an outer window, which are inclined to one another. The angle of inclination was found by measurement of the blueprints to be about 6°. Light passing through the outer window and reflected by the inner one, back to the outer window, and then back again into the spacecraft would have been turned through an angle of about 12°, in a direction away from the top of the spacecraft, which in flight, points near the horizon. This explanation probably accounts for what was seen in the photographs. The existence of these reflections has been directly verified in the Mercury procedures trainer at the Mercury Control Center. It was further found in spacecraft 18 that one of the reflections (there were two) had a light tan color corresponding to that observed by Glenn.

Since it is a spacecraft phenomenon, the luminous band produced by reflection must also have been present in the night sky, especially after moonrise. It may be identical with the band observed by Glenn. The color difference which he remarked on may have resulted from an anti-reflectant coating which had been applied to the windows.

If not due to reflection, it might be possible to attribute the band to some auroral phenomenon. There is a line in the auroral spectrum at 5,577 Å. This line is known from rocket measurements to stop at 100 kilometers. A height of 100 kilometers would appear, at the spacecraft height of 250 kilometers, as a false horizon at an angular altitude of about 3°. It would be green in color, and would be more difficult to see after moonrise. It does not agree with the luminous band.

In addition, there are two auroral red lines, at 6,300 and 6,464 Å, which are known to come from a height greater than any so far reached by rockets sent to observe them. From theory, they ought to be at a height of about 240 kilometers. These might be reconciled with the observed luminous band, though they ought not to be easier to see after moonrise. They would explain the tan to buff color observed. On the other hand, these lines are much fainter than 5,577, so that it is hard to understand why they would be observed while it was missed.

On the whole, the balance of probability is that the luminous band was due to reflection in the spacecraft window. The outstanding reason for connecting the two is that the inclined windows should have given a ghost image.

The Flattened Sun

Glenn reports that the sunset appeared to be normal until the last moment, when the sun appeared to spread out about 10° on either side, and to merge with the twilight band. He specifically states that he did not see the sun as a narrow, flat object.

On the other hand, three consecutive photographs of the setting sun can be well interpreted in terms of the theoretically predicted sausage shape. In two of these, there is some slight spreading of the image, evidently partly photographic and partly due to motion; and in the third, the motion is considerable. All, however, appear to indicate a solar image about ½ degree in greatest dimension as required by theory, rather than a much shorter length, as would be found if the setting were like that seen from the ground.

The Ultraviolet Photography

A total of six spectrograms were taken of the Orion region. All six show the stars of the Belt; most also show Rigel and the Sword stars. One, more carefully guided than the others, reaches magnitude 4.0 and includes 14 images, some underexposed, and Rigel overexposed. In all cases, it appears from the (very approximately) uniform light distribution that the spacecraft was yawing at a steady rate.

The pictures have not yet been correlated with the spacecraft program. It is understood that they represent exposures of roughly 15 seconds.

Standardization in the blue region was accomplished when the pictures were developed. Ultraviolet standardization will be done at Eastman Kodak.

Next, it is planned to determine the long and short wavelength cutoffs, from spectra exposed at McDonnell Aircraft Corporation through the spacecraft window and to find the dispersion by further studies,

Finally, it is hoped to construct curves giving the relative intensities in the stars observed over the available region, and to compare with the work of Kupperian at NASA Goddard Research Center.

(None of these plans can be considered official or final at this time.)

Reference

1. MINNAERT, M., (H. M. Kremer-Priest and K. E. Brian Jay, trans.) : Light and Color in the Open Air. Dover Publications, c. 1954, p. 136.

MERCURY SPACECRAFT INTERIOR ARRANGEMENT
from Project Mercury Indoctrination
revision May 21 1959

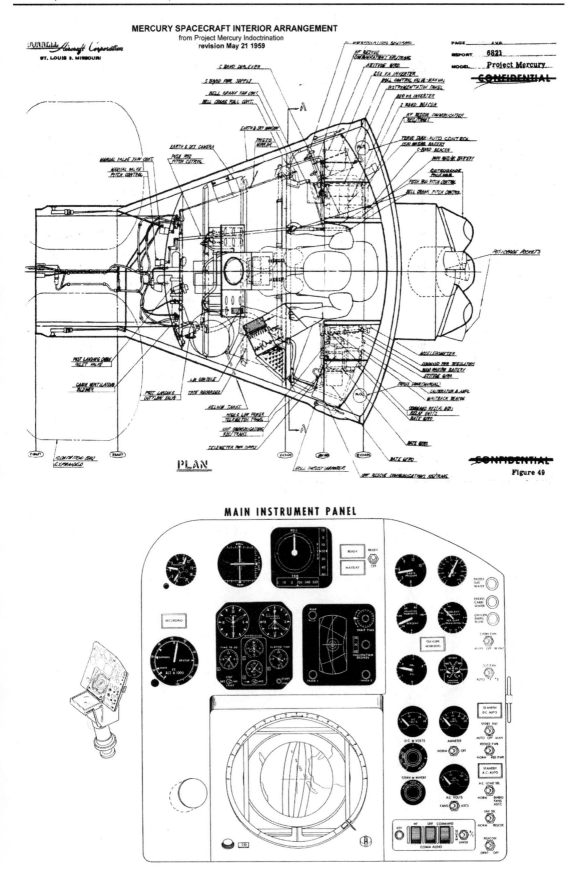

Figure 49

MAIN INSTRUMENT PANEL

CONSOLE PANELS

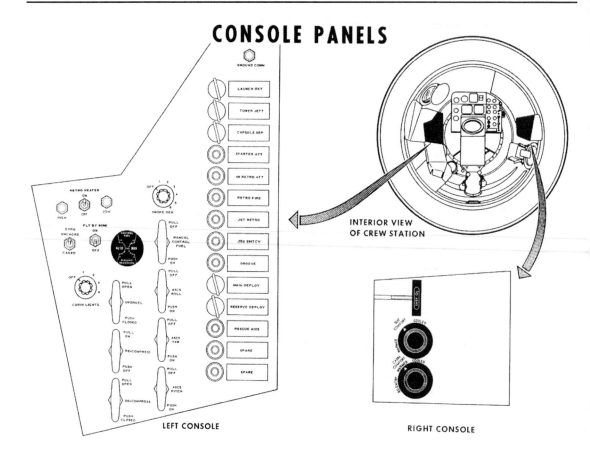

LEFT CONSOLE

INTERIOR VIEW
OF CREW STATION

RIGHT CONSOLE

THREE AXIS HAND CONTROL

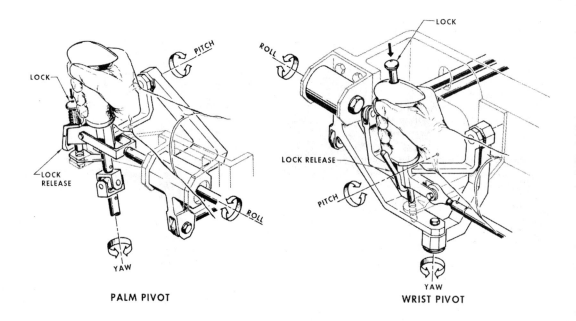

PALM PIVOT

WRIST PIVOT

NAVIGATIONAL AIDS
(SCOPE DISPLAY, EARTH PATH INDICATOR AND CHART BOARD)

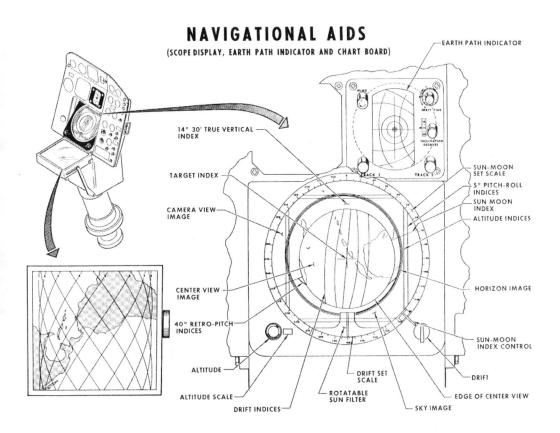

EARTH PATH INDICATOR

14° 30' TRUE VERTICAL INDEX

TARGET INDEX

CAMERA VIEW IMAGE

CENTER VIEW IMAGE

40° RETRO-PITCH INDICES

ALTITUDE

ALTITUDE SCALE

DRIFT INDICES

ROTATABLE SUN FILTER

DRIFT SET SCALE

SKY IMAGE

EDGE OF CENTER VIEW

DRIFT

SUN-MOON INDEX CONTROL

HORIZON IMAGE

ALTITUDE INDICES

SUN MOON INDEX

5° PITCH-ROLL INDICES

SUN-MOON SET SCALE

John Glenn explains the details of his flight to President Kennedy at the Cape Canaveral facility. (below)

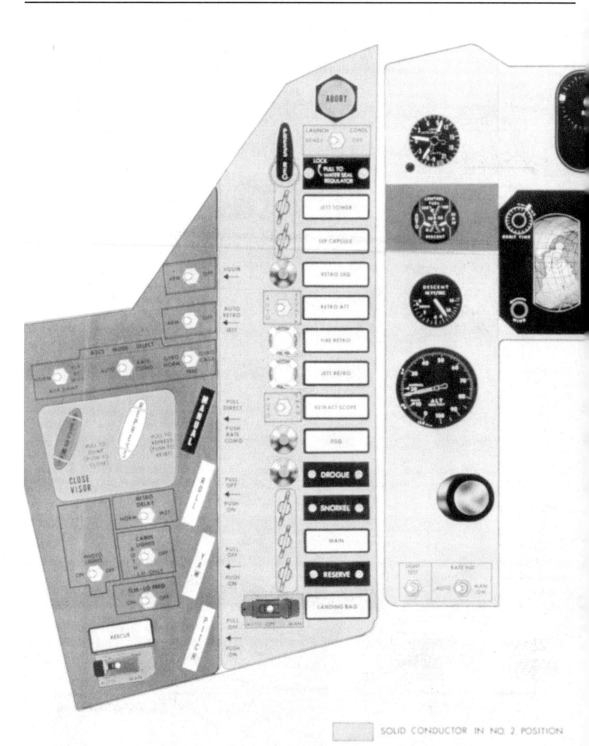

SOLID CONDUCTOR IN NO. 2 POSITION

The Main Control Panel of Friendship Seven (above)

John Glenn suits up ready for the flight, note the large circular mirror on his chest. (below)

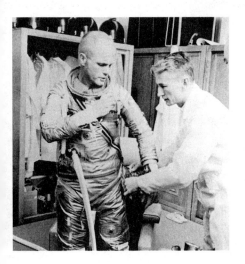

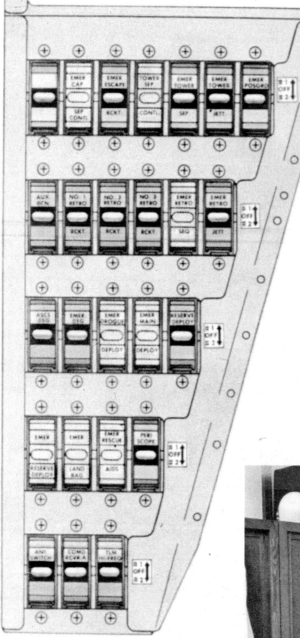

Auxiliary Control Panel showing Retro, escape tower, parachute, periscope & recovery controls (left)

John Glenn ready for the history books.